Investigation and Monetary Values of Fish and Freshwater Mollusk Kills

Investigation and Monetary Values of Fish and Freshwater Mollusk Kills

Edited by

Robert I. Southwick
Southwick Associates
Post Office Box 6435
Fernandina Beach, Florida 32035, USA

Andrew J. Loftus
3116 Munz Drive, Suite A
Annapolis, Maryland 21403, USA

American Fisheries Society
Special Publication 35

The American Fisheries Society Special Publication series is a registered serial. A suggested citation format for this book follows

Southwick, R. I., and A. J. Loftus, editors. 2017. Investigation and monetary values of fish and freshwater mollusk kills. American Fisheries Society, Special Publication 35, Bethesda, Maryland.

Cover images: top left, U.S. Fish and Wildlife Service; top right, courtesy of the Missouri Department of Conservation; bottom, Janet L. Clayton, West Virginia Division of Natural Resources.

Second printing, August 2018, to revise Appendix A.

Printed in the United States of America on acid-free paper.

Library of Congress Control Number 2017955455

ISBN 978-1-934874-47-9
ISSN 0097-0638

American Fisheries Society Web site: www.fisheries.org

American Fisheries Society
425 Barlow Place, Suite 110
Bethesda, Maryland 20814
USA

Contents

Preface

This publication, *Investigation and Monetary Values of Fish and Freshwater Mollusk Kills*, has been prepared for natural resource professionals who investigate fish and freshwater mussel kills and assess monetary damages of such events. Previous editions and predecessors of this publication have been adopted as the legal basis for fines or restitution in more than half the states and have been upheld in numerous legal challenges. This manual contains guidelines for organizing investigations, designing and performing necessary fieldwork, and calculating legally defensible damages to fish or freshwater mussel populations. See page xv, the first pages of Chapters 4 and 6, and periodic footnotes throughout the document for important limitations on the use of the economic guidelines.

This book represents continuing efforts of the members of the American Fisheries Society (AFS) who have worked for more than 47 years to develop and refine methods for sampling and estimating the magnitude and economic value of fish kills. During the preparation of the 2003 and 2017 editions, AFS was joined by the Freshwater Mollusk Conservation Society (FMCS) in publishing guidelines for assessing and assigning economic damages to freshwater mussel kills.

Chapter 1 ("Procedures for Investigating a Fish Kill"), a set of guidelines for coordinating and documenting a field investigation, is a revised version of *Fish Kill Investigation Procedures*, originally published by the AFS Southern Division's Pollution Committee in 1986. Chapter 2 ("Principles of Sampling Dead Fish") and Chapter 3 ("Field Guidelines for Counting Dead Fish") are the latest expression of the Pollution Committee's work in developing standard procedures for sampling fish kills. The Pollution Committee's *Fish Kill Counting Procedures* was first published in 1970, revised in 1979 (as *Fish Kill Counting Guidelines*), and revised again in 1982, 1992, and 2003. The 1982 revision was Part II of *Monetary Values of Freshwater Fish and Fish Kill Counting Guidelines,* AFS Special Publication 13; the 1992 revision was part of *Investigation and Valuation of Fish Kills*, AFS Special Publication 24; and the 2003 revision was part of *Investigation and Monetary Values of Fish and Freshwater Mussel Kills,* AFS Special Publication 30 (Southwick and Loftus 2003).

Chapter 4 ("Costs and Economic Valuation of Fish Kills") dates back to the Pollution Committee's *Monetary Values of Fish* booklets of 1970 and 1975, which dealt with southern U.S. species. In 1978, the AFS North Central Division's Monetary Values of Fish Committee published *Reimbursement Values for Fish*, addressing species in 12 northern states and two Canadian provinces. To integrate these and other regional values, a special AFS Monetary Values of Freshwater Fish Committee collected values from 135 federal, state, provincial, and private agencies and hatcheries. These data were published in 1982 as Part I of AFS Special Publication 13. For the 1992 publication of *Investigation and Valuation of Fish Kills*, the 2003 publication of *Investigation and Monetary Values of Fish and Freshwater Mussel Kills*, and the present book, under guidance of the Socioeconomics Section, the survey was repeated to update replacement costs for killed fish and to summarize procedures for estimating the broader economic losses resulting from a fish kill.

Chapter 5 (" Procedures for the Investigation of Freshwater Mollusk Kills") and Chapter 6 ("Costs and Economic Valuation of Freshwater Mussel Kills") were modeled after the procedures and format outlined for the fish kill section and thereby take advantage of the history and precedence established in these sections during the past 47 years. This publication represents the second iteration of this type of information for freshwater mussels in such a comprehensive manner and reflects updated information resulting from advancements in understanding of freshwater mussel life history, ecology, and culture since its initial publication in 2003. These chapters were developed primarily by the FMCS, with editorial oversight by AFS. The survey methodology and implementation were developed under close oversight by, and involvement of, a steering committee composed of the members of the FMCS.

We fully expect that *Investigation and Monetary Values of Fish and Freshwater Mollusk Kills* will be revised as more data are collected regarding fish, freshwater mussels, and potentially additional aquatic flora and fauna and as the understanding of damage assessment develops. Again, we urge that fishery biologists and managers apply their professional judgments and expertise when using this book.

Robert I. Southwick and Andrew J. Loftus, editors

Acknowledgments

This document represents the collective effort of many individuals. Three groups spearheaded its development and provided leadership: the Pollution Committee of the American Fisheries Society's (AFS) Southern Division, the AFS Socioeconomics Section, and the Freshwater Mollusk Conservation Society (FMCS). Funding, guidance, and other support were provided by the U.S. Fish and Wildlife Service Fish and Aquatic Conservation Program under award F15AP00972, the U.S. Fish and Wildlife Service Midwest Region Fisheries Program under award F16AP00589 (Genoa National Fish Hatchery), FMCS, the Texas Parks & Wildlife Department, the Indiana Department of Natural Resources Division of Fish and Wildlife, the Missouri Department of Conservation, the New York State Department of Environmental Conservation, the Ohio River Natural Resources Damage Assessment Trustee Council, the Arizona Fish and Game Department, the Georgia Department of Natural Resources, the Alabama Department of Conservation and Natural Resources Division of Wildlife and Freshwater Fisheries, and Upper Mississippi River Conservation Committee, and AFS Southern Division.

This report was edited by Southwick Associates, Inc. of Fernandina Beach, Florida and Andrew J. Loftus of Annapolis, Maryland. Tom Allen led efforts at Southwick Associates with assistance from Dr. Kristie Blevins of Eastern Kentucky University. The sections on fish kill investigation procedures (Chapters 1–3) were originally written by the AFS Southern Division Pollution Committee for the 1992 document. For this revision, these chapters were reviewed and updated by the same committee under the guidance of Rebecca O'Hearn (Missouri Department of Conservation) and Bryant Bowen (Georgia Department of Natural Resources). Chapter 5 ("Procedures for the Investigation of Freshwater Mollusk Kills") and Chapter 6 ("Costs and Economic Valuation of Mussel Kills") were originally developed in 2003 by the FMCS under the guidance of John Van Hassel of American Electric Power. For this publication, these chapters have been extensively reviewed by FMCS under the leadership of Megan Bradley, Janet Clayton, Heidi Dunn, Rachael Hoch, Patricia Morrison, Teresa Newton, and Jeremy Tiemann.

This report would not have been possible without the significant contributions of many, including members and leadership of the AFS Southern Division's Pollution Committee led by Rebecca O'Hearn and Bryant Bowen, the AFS Socioeconomics Section, led by Tom Lang; the members and leadership of FMCS, and AFS advisory panel members who guided this effort. Contributors include Scott Morrison, Dan Hill, Kelly Winningham, Frank Borsuk, Dr. George Guillen, Don Pitts, M. Travis Tidwell, Kathryn Burger, Sarah Haas, Greg Conley, Lindsey Saum, Angela Schrift, and Johanna Gregory.

The editors express their appreciation to the personnel from the fish hatcheries and mussel production facilities who completed surveys and took time to respond to follow-up questions. The editors also wish to thank the AFS Southern Division's Pollution Committee for their time, input, and recommendations; the numerous reviewers of the initial draft; and the staff of the American Fisheries Society (Doug Austen, Dan Cassidy, Aaron

Lerner, and Debby Lehman). There are many more people who contributed considerable time and data to this project—too many to list—who deserve recognition and thanks. Regardless of the contributions from the many individuals and organizations listed here, the editors remain responsible for all content reported herein.

Symbols and Abbreviations

The following symbols and abbreviations may be found in this book without definition. Also undefined are standard mathematical and statistical symbols given in most diction-

A	ampere
AC	alternating current
Bq	becquerel
C	coulomb
°C	degrees Celsius
cal	calorie
cd	candela
cm	centimeter
Co.	Company
Corp.	Corporation
cov	covariance
DC	direct current; District of Columbia
D	dextro (as a prefix)
d	day
d	dextrorotatory
df	degrees of freedom
dL	deciliter
E	east
E	expected value
e	base of natural logarithm (2.71828...)
e.g.	(exempli gratia) for example
eq	equivalent
et al.	(et alii) and others
etc.	et cetera
eV	electron volt
F	filial generation; Farad
°F	degrees Fahrenheit
fc	footcandle (0.0929 lx)
ft	foot (30.5 cm)
ft^3/s	cubic feet per second (0.0283 m^3/s)
g	gram
G	giga (10^9, as a prefix)
gal	gallon (3.79 L)
Gy	gray
h	hour
ha	hectare (2.47 acres)
hp	horsepower (746 W)
Hz	hertz
in	inch (2.54 cm)
Inc.	Incorporated
i.e.	(id est) that is
IU	international unit
J	joule
K	Kelvin (degrees above absolute zero)
k	kilo (10^3, as a prefix)
kg	kilogram
km	kilometer
l	levorotatory
L	levo (as a prefix)
L	liter (0.264 gal, 1.06 qt)
lb	pound (0.454 kg, 454g)
lm	lumen
log	logarithm
Ltd.	Limited
M	mega (10^6, as a prefix); molar (as a suffix or by itself)
m	meter (as a suffix or by itself); milli (10^{-3}, as a prefix)
mi	mile (1.61 km)
min	minute
mol	mole
N	normal (for chemistry); north (for geography); newton
N	sample size
NS	not significant
n	ploidy; nanno (10^{-9}, as a prefix)
o	ortho (as a chemical prefix)
oz	ounce (28.4 g)
P	probability
p	para (as a chemical prefix)

p	pico (10^{-12}, as a prefix)
Pa	pascal
pH	negative log of hydrogen ion activity
ppm	parts per million
qt	quart (0.946 L)
R	multiple correlation or regression coefficient
r	simple correlation or regression coefficient
rad	radian
S	siemens (for electrical conductance); south (for geography)
SD	standard deviation
SE	standard error
s	second
T	tesla
tris	tris(hydroxymethyl)-aminomethane (a buffer)
UK	United Kingdom
U.S.	United States (adjective)
USA	United States of America (noun)
V	volt
V, Var	variance (population)
var	variance (sample)
W	watt (for power); west (for geography)
Wb	weber
yd	yard (0.914 m, 91.4 cm)
α	probability of type I error (false rejection of null hypothesis)
β	probability of type II error (false acceptance of null hypothesis)
Ω	ohm
μ	micro (10^{-6}, as a prefix)
′	minute (angular)
″	second (angular)
°	degree (temperature as a prefix, angular as a suffix)
%	per cent (per hundred)
‰	per mille (per thousand)

Important: Read This First

The methods and economic data in this book are guidelines only. They are not intended to supercede existing state, provincial, or federal methods for estimating damages after a fish or freshwater mussel kill. The American Fisheries Society strongly recommends that those investigating fish and mussel kills use professional judgment and expertise to conduct specific studies to adjust the economic values herein whenever these steps are needed to reflect local conditions. The information in this book is insufficient for evaluating monetary damages from the following types of kills:

- Large-scale kills causing loss of recreational fisheries or damage to large, heavily used bodies of water;
- Kills occurring in unique aquatic habitats;
- Kills of species protected by local, state, or provincial regulations or listed as threatened or endangered under federal statutes;
- Kills of trophy-sized fish.

For additional restrictions, consult the statements of applicability in Chapters 4 and 6.

Specialized studies are needed to estimate damages for such kills. If circumstances prevent such studies, procedures in Chapters 4 and 6 may be used as proxies for better estimates of value. These procedures, however, may result in underestimation of damages.

Readers should heed the various guidance on uses of the information that are included in this document, particularly pertaining to freshwater mussels, threatened and endangered species, and special circumstances.

Chapter 1
Procedures for Investigating a Fish Kill

Investigation of a fish kill can involve several state, provincial, and federal agencies and may require the expertise of professionals from several disciplines. Cooperation among these agencies and professionals is essential for the thorough investigation and successful prosecution of any case. One agency should have the lead responsibility for fish kill investigations. Other state, provincial, and federal agencies that investigate fish kills or water pollution should carefully define their legal authority and areas of expertise and be prepared to assist the lead agency within these limits. Because federal, state, and provincial responsibilities can overlap, a standard notification procedure should be adopted by all the agencies involved.

The responsibility for investigating a fish kill can rest with state, provincial, tribal, and/or federal governments, and frequently, responsibilities are shared. In the United States, for example, the Coast Guard shares responsibility with the U.S. Environmental Protection Agency and state environmental agencies for investigation of spills of oil or designated hazardous substances into navigable waters. At the federal level, the U.S. Fish and Wildlife Service (USFWS) generally has jurisdiction if freshwater fish are killed, although this agency often delegates its authority to state government. The National Marine Fisheries Service (NMFS) may also have authority in the investigation if marine or anadromous fish are killed or if the incident occurs in federal offshore waters. Various state agencies may have important investigative roles (Table 1.1). For a summary of agency responsibilities, see Hill (1983); for natural resource damage assessment regulations established by the U.S. Department of Interior, including USFWS, and used by trustees of natural resources for spills of hazardous substances, see USOFR (1996); for damage assessment regulations established by the National Oceanic and Atmospheric Administration and used by trustees of natural resources for discharges of oil, see USOFR (1990). Cooperation among these agencies and professionals is essential for the thorough investigation and successful prosecution of any case.

The U.S. Endangered Species Act gives the USFWS and NMFS the authority to investigate fish kills if endangered or threatened fish are harmed or if the affected area is critical habitat for such species. For endangered as well as other species, the USFWS usually has responsibility for freshwater fish and NMFS for marine and anadromous fish. In Canada, the federal Minister of Fisheries has authority to investigate fish kills caused by pollution; however, in some provinces, this authority is shared with the provincial government. Several provinces also have legislation protecting rare and endangered species.

Initial Notification

When notified about a fish kill, the lead agency's representative should gather as much pertinent information as possible from the reporting party. The fish kill notification form

TABLE 1.1. Examples of typical responsibilities of state agencies involved in fish kill investigations.

Agency type	Investigative role or contribution
Fisheries management	Provides fisheries expertise. Estimates numbers and species of fish killed. Estimates valuation of kill plus investigative costs for a total investigative cost number. Plans and implements restoration measures.
Water quality control	Provides limnological and oceanographic experts and qualified chemists. Uses chemical and physical measurements to delineate the affected area and document the source and type of pollution. May serve as liaison with federal agencies and other emergency response teams.
Environmental protection	Provides water quality experts and legal counsel. Usually has authority to regulate water quality and, consequently, has an active role in the investigations, often as the state on-scene coordinator.

is a checklist of the basic information needed for a well-documented report (Figure 1.1). This form should be readily available at all offices involved in fish kill investigations. The name, address, and phone number of the person reporting the kill should be recorded for use at a later date. The exact location and time of the kill are needed to locate the affected area and determine the type of equipment needed to investigate the kill. Information on the species of fish involved, condition of the fish (whether or not they are still dying), and possible sources of pollution can be useful. An opinion from the reporting party about the cause of the kill will also help in the investigation.

Coordinating the Response Effort

In most cases, personnel from a provincial or state pollution response team will be notified first about a fish kill. However, it is not uncommon for other agencies to receive the first notification. Each agency should have clearly written guidelines/standard operating procedures, based on agency policy, for the dissemination of fish kill information.

The guidelines may instruct the individual receiving the initial report to notify other agencies about the incident. These agencies may include any federal, state, provincial, or local agency concerned with environmental protection, water quality, navigable waters, fish and other aquatic resources, disaster and emergency responses, and public health and safety. This notice opens communications between all interested parties and gives the investigating team access to many types of expertise. For example, district fisheries biologists and state or provincial employees may be excellent sources of information about a kill, especially if the pollution source is not readily apparent. Notification does not necessarily result in additional support from other agencies for field investigations.

Consultation with the pollution response agency to discuss health hazards on-scene is important prior to responding. If conditions are hazardous to human health, proper per-

FISH KILL NOTIFICATION

STATE: ______________________

DATE OF KILL: ________________

DATE REPORTED: ________________ TIME REPORTED: __________

NAME OF REPORTER: ______________________________

ADDRESS: ______________________ PHONE: __________

ORGANIZATION: ______________________________

WATER(s) INVOLVED: ________________ COUNTY: __________

SPECIFIC LOCATION OF KILL (bridge, highway or state road, industry, landmark, mountain, park,etc.): ______________________________

____

____

Township: ______ Range: ______ Section: ______ 1/4 Section: ______

SUSPECTED REASON FOR FISH KILL: ______________________

____

LOCATION OF SOURCE: ______________________________

NAME OF ALLEGED POLLUTER: ______________________

ADDRESS: ______________________ PHONE: __________

SPECIES INVOLVED: ______________________________

NUMBER: ________________ FISH STILL DYING? __________

PERSONS AND AGENCIES NOTIFIED:

	NAME	TIME	PHONE
1.			
2.			
3.			

ADDITIONAL COMMENTS: ______________________________

____

____

REPORT RECEIVED BY

____

FIGURE 1.1. Example of an initial notification form.

sonal protective equipment will need to be utilized, or if unavailable, the investigator will need to wait for safety clearance.

Assembling Materials and Equipment

The key to a successful investigation is advance preparation for any field procedures that might be needed at the kill site. Each agency should develop a set of field procedures for conducting a fish kill investigation. Gathering as much information on the fish kill prior to going in the field is essential for proper planning and preparation. An equipment checklist is nearly as important as the equipment itself (Table 1.2). (Do not try to memorize all items needed for a thorough investigation!) The best approach is to keep a basic field equipment kit ready at all times.

On-Site Procedures

After arriving at the site of a fish kill, the pollution response team has two types of activities: reconnaissance and documentation. Reconnaissance should be carried out along with but not before documentation.

Reconnaissance

During reconnaissance, the response team tries to determine the source of the pollution and takes action to stop it (if this has not already been done). These actions are essential to minimize damage, but they should not delay documentation of the fish kill.

The person or company suspected of having caused the kill should be notified immediately. Prompt notification protects the investigating agency from accusations of secrecy and allows alleged offenders to conduct their own investigation as soon as possible.

Documentation

The response team should begin documenting the fish kill immediately upon arrival. The activities described below can be carried out concurrently by various team members. Decisions on what types of sampling and water testing to conduct should be based on knowledge of potential pollutants or causes.

Water sampling.—Water samples should be taken from sites inside and outside of the study area (area encompassing the entire fish kill) and from any tributaries that enter the affected body of water. Particular emphasis should be placed on collecting control samples from nonimpacted sites for a baseline comparison. Multiple samples collected above, within, and below the kill zone not only delineate the zone, they also may pinpoint the pollution source or document that a chronic problem exists.

Water samples should be representative of the polluted water suspected to have caused the fish kill. Ideally, samples should be integrated over depth and width. Alternatively, multiple samples may be collected across the extent of the area and at various depths to adequately characterize the water body. Response team members should be warned, if they do not already know, that grab samples may be unrepresentative, particularly if the water body is not evenly mixed or if the pollutant is suspended rather than dissolved.

TABLE 1.2. Checklist of equipment for investigating a fish kill.

Necessary equipment

Weighing scales
Measuring board
Cooler with ice
Plastic bags with labels and fasteners
Water sample containers (glass, plastic)
Camera, film or digital with extra batteries and memory card
Global Positioning System with extra batteries
Area map
List of cooperating agencies and phone numbers
Preservatives (formalin, alcohol, nitric and sulfuric acids)
Aluminum foil
Evidence tags
Field notebook for chronological log
Clipboards for waterproof forms
Waterproof forms:
  Fish kill notification form
  Chain of custody forms
  Fish kill investigation report form
  Fish-counting forms
Pencils and waterproof marking pens
Appropriate personal protective equipment, such as gloves and boots
First aid kit
Shovel, tubs, buckets
Dip nets
100-yd measuring tape, range finder
Flagging
Water test meters and kits (calibrated prior to daily use)
Tally meters or handheld clicker
Calculator
Flashlights and spare batteries
Procedural checklist, manual, or on-scene interpretive keys
Trash bags

Optional equipment
(some may be essential for lake, estuary, or marine kills)

Boat, life vest (personal flotation device), and related boating equipment
Audio and video recorders
Two-way radio, cell phone, etc., chargers, and spare batteries.
Rope
Stopwatch
Secchi disk
Compass

TABLE 1.2. Continued.

Optical distance meter
Pitch fork for large numbers of fish
Plankton net
Dissecting pan and tools

Note: Exclusive reliance on battery-powered instruments is not advisable. Other methods of analysis should also be available in case instruments fail. For example, materials for a Winkler analysis should be available in case an oxygen meter malfunctions.

Water samples should not be contaminated with bottom sediments suspended by team activities. Therefore, water sampling should be done in a way that minimizes bottom disturbances, and it should be completed before fish, invertebrates, and other samples are collected.

When laboratory analysis will be required for suspected pesticides, heavy metals, and other chemicals, duplicate water samples should be taken at each sample site for quality control purposes. Plastic containers normally are used for samples destined for metals analysis. Acid-washed glass bottles should be used for samples to be analyzed for organic contaminants (pesticides, petroleum products, etc.) and nutrients. Standard techniques should be used to collect and preserve water samples. The latest edition of *Standard Methods for the Examination of Water and Wastewater* (APHA et al. 2012 is the 22nd edition) contains generally accepted procedures. In the United States, workers also should follow sampling protocols currently approved by the U.S. Environmental Protection Agency, most of which are adopted from *Standard Methods for the Examination of Water and Wastewater.*

Water testing.—Water quality variables usually measured at kill sites include water temperature, dissolved oxygen, pH, total ammonia, and conductivity. All instruments should be calibrated before and after use. Calibration data should be logged to ensure that measurements of water quality can be defended in court.

Examination of fish and macroinvertebrates.—Dead or dying fish and benthic macroinvertebrates should be collected from the kill site. Fish should be examined externally for gross abnormalities, such as hemorrhaging, open lesions, and discoloration. If appropriate, fish and invertebrates should be collected for tissue analysis or parasite examination. Sampled animals should be collected as fresh as possible, placed in bags or containers, and placed on ice; those collected for chemical analyses should be frozen as soon as possible. The behavior of stressed fish should be noted because loss of equilibrium, listlessness, nervousness, or convulsive action may indicate a specific pollutant (Table 1.3). The *Field Manual for the Investigation of Fish Kills* (Meyer and Barclay 1990) is an excellent source of standard procedures for field examinations of dead fish.

Fish counts.—Counts of dead fish should be made according to procedures outlined in Chapters 2 and 3 of this volume (for an example of a fish-counting form, see Figure 1.2). These counts will be used to assess damages for the killed fish.

Record keeping.—The leader of the response team should keep a chronological log of events during the investigation. The log should include the names of people interviewed,

TABLE 1.3. Conditions that may be observed at a fish kill site and their possible causes.

Condition	Possible cause
Fish come to the surface (usually larger fish) and gulp air and the water has low levels of dissolved oxygen.	Organic matter has depleted the dissolved oxygen. Look for a sewage treatment plant or lift station, livestock feedlot, irrigation runoff, or algal bloom (green water). Did a large rain event occur the night previous of the kill?
Fish come to the surface and gulp air but the water has adequate levels of dissolved oxygen.	Dissolved oxygen may have been depleted but partly restored over time; look for the source of organic matter listed above. Ammonia may have reached toxic concentrations; check water chemistry and look for a livestock feedlot. Check for an infestation of gillworms; for a toxic algal bloom if the water is discolored; and for a source of chlorine if scales or gills are bloody and a sewage treatment plant or municipal water lines are nearby.
Fish are dying after a heavy rain.	Pesticides or other chemicals may have been washed off adjacent agricultural fields, dumped from spraying equipment, or discharged during aerial spraying. Toxic or septic sediments could have been stirred up. Runoff of organic material and nutrients may cause an algal bloom and low levels of dissolved oxygen. Rainfall can also break down the thermocline and create turnover by causing the hypoxic bottom water to be distributed throughout the water column.
Water has an oily sheen.	A drilling operation, a refinery, or a leaky pipeline is in the area, or petroleum spilled from a truck or service station.
Streambanks and stream bottom are covered with an orange substance and water conductivity is high.	Oil drilling operations have discharged brine water.
Water has low pH, with or without orange discoloration of substrate but good water clarity.	Acidic water has discharged from a coal mine or a chemical spill has occured.
Mostly small fish, but not always exclusively, along the shoreline have died and air temperature is subfreezing.	Fish were killed by excessive cold.
Mostly small fish, but not always exclusively, are dead below a dam or an industrial plant that discharges heated water.	Fish were killed by passage through dam gates or hydroelectric turbines, or from thermal shock.
Kill is restricted to one species or size-class.	Fish may have died from spawning stress, disease pathogens, or schooling in poor water quality.

notes from the interviews, water-sampling and fish-counting locations, and other details of the investigation. This log can be used later to prepare the fish kill report, identify omissions in the investigation process, document field activities for legal proceedings, and determine the cost of the investigation (including salaries, travel expenses, photographic supplies, analyses of samples, and administrative time) if the involved agencies will be seeking monetary damages.

All samples should be completely identified with evidence tags (Figure 1.3), and the chain of custody should be preserved for all samples throughout the analysis (Figure 1.4). Poorly identified samples or a break in the chain of custody can jeopardize the prosecution of a pollution case.

Photographs and well-prepared maps should be used extensively for documentation. The time, date, location, subject, and intent of each photograph, as well as the photographer's name, should be recorded in a notebook. Photographs may be used to document pollution sources, discharge points, water discoloration, types of fish killed, magnitude of the kill, and corrective action taken. These photographs are useful references and can be very influential during court proceedings.

Public Relations

Fish kills are news and reporters often arrive while an on-site investigation is in progress. The lead agency should designate one person as the spokesperson. This person should provide facts about the kill but not make guesses or offer opinions about the cause of the kill. Opinions and guesses are unfair to the potentially responsible party and may hinder future legal actions. The public may need an informed statement about fish kills. For example, it may be beneficial for a public statement to be made about the portion of dead fish that are actually observed during an event or that different species and sizes of fish react differently to pollutants.

Report Preparation

The lead agency often prepares the fish kill investigation report, which documents in detail all aspects of the investigation (Figure 1.5). Accuracy and completeness are crucial because this report may undergo judicial scrutiny, and omissions and oversights can render a high-quality investigation useless.

All reports should be prepared as if the case were to be taken to court. All data should be included and professionally presented. A detailed and concise report will help explain the case to the potentially responsible party. In addition, attorneys for the investigating agency can use the report to prepare their case.

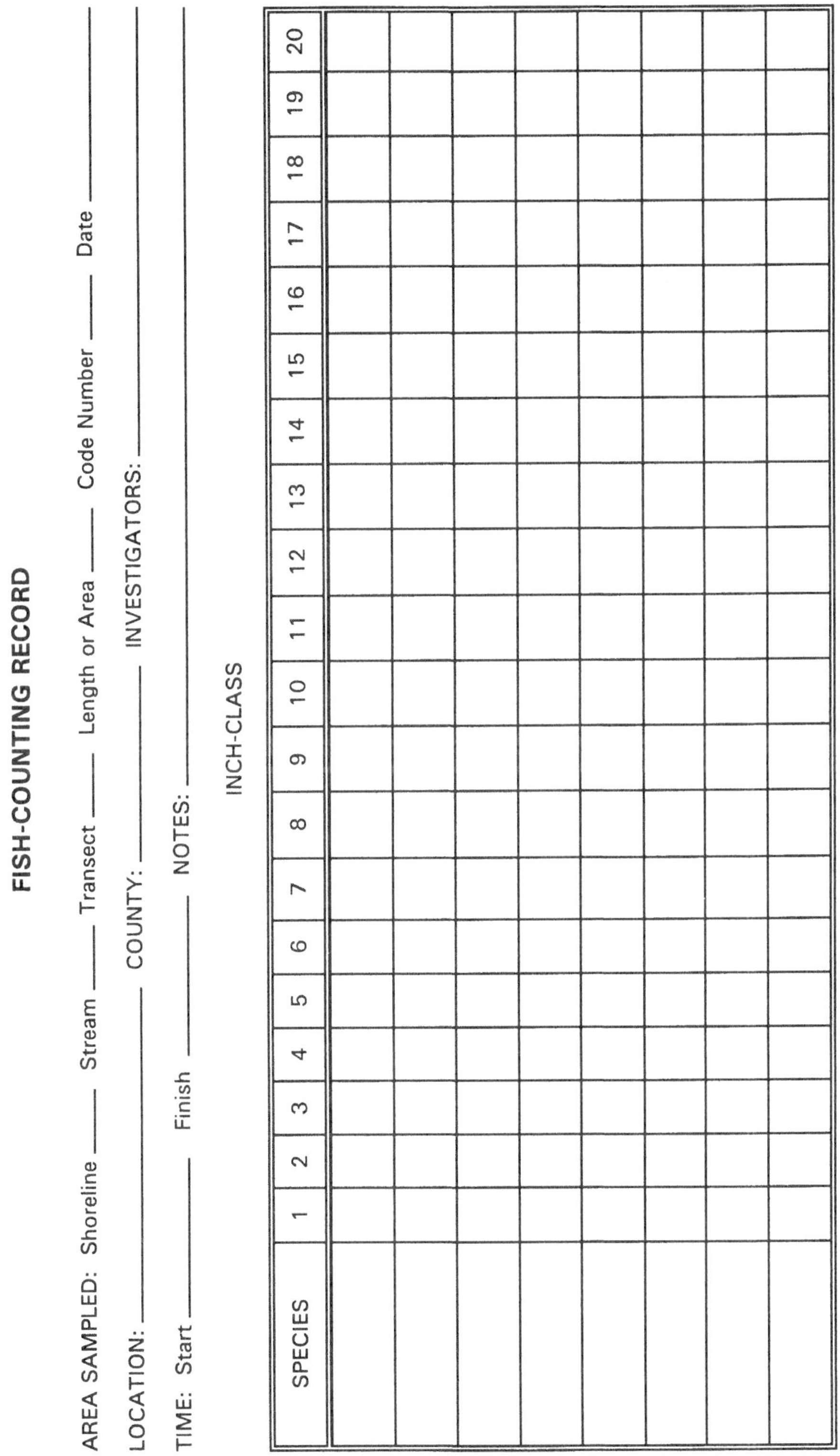

FISH-COUNTING RECORD

AREA SAMPLED: Shoreline ______ Stream ______ Transect ______ Length or Area ______ Code Number ______ Date ______

LOCATION: ______ COUNTY: ______ INVESTIGATORS: ______

TIME: Start ______ Finish ______ NOTES: ______

INCH-CLASS

SPECIES	1	2	3	4	5	6	7	8	9	10	11	12	13	14	15	16	17	18	19	20

FIGURE 1.2. Example of a fish kill counting form.

FRONT

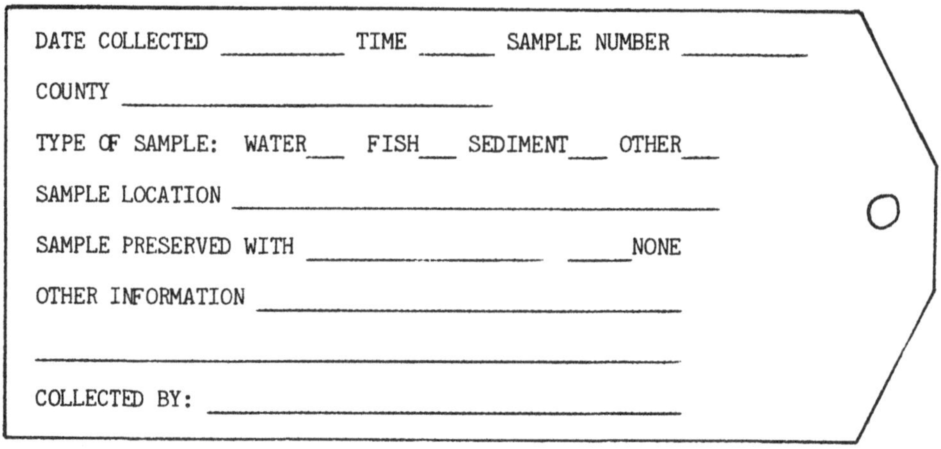

DATE COLLECTED ________ TIME _____ SAMPLE NUMBER ________

COUNTY ____________________

TYPE OF SAMPLE: WATER___ FISH___ SEDIMENT___ OTHER___

SAMPLE LOCATION ____________________

SAMPLE PRESERVED WITH ____________ _____NONE

OTHER INFORMATION ____________________

____

COLLECTED BY: ____________________

BACK

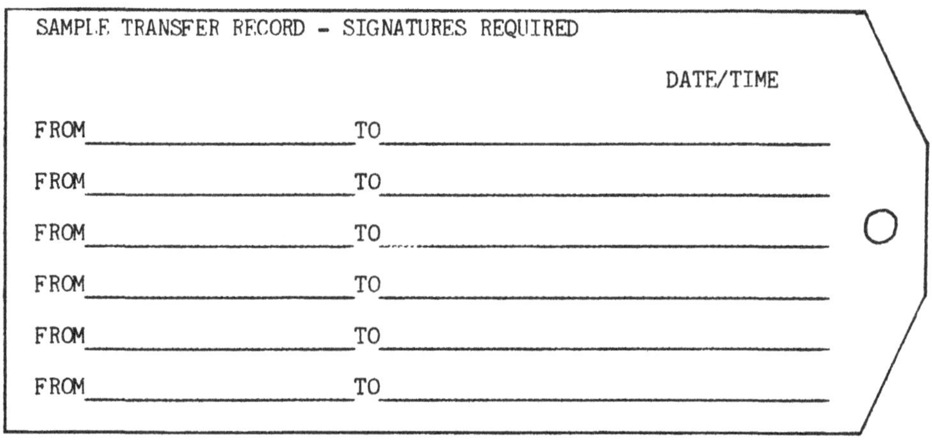

SAMPLE TRANSFER RECORD - SIGNATURES REQUIRED

DATE/TIME

FROM__________TO__________

FROM__________TO__________

FROM__________TO__________

FROM__________TO__________

FROM__________TO__________

FROM__________TO__________

FIGURE 1.3. Example of an evidence tag.

Suggested Reference Material

Several reference works, in addition to the ones mentioned in the text, contain information useful for planning and conducting a fish kill investigation and for preparing an investigative report. The documents listed in Table 1.4 constitute a good basic library for those involved with fish kill investigations. **Important: Publications are continually updated, and often published online or electronically. Investigators should consult the publisher of any work for the most recent updated versions that are available.**

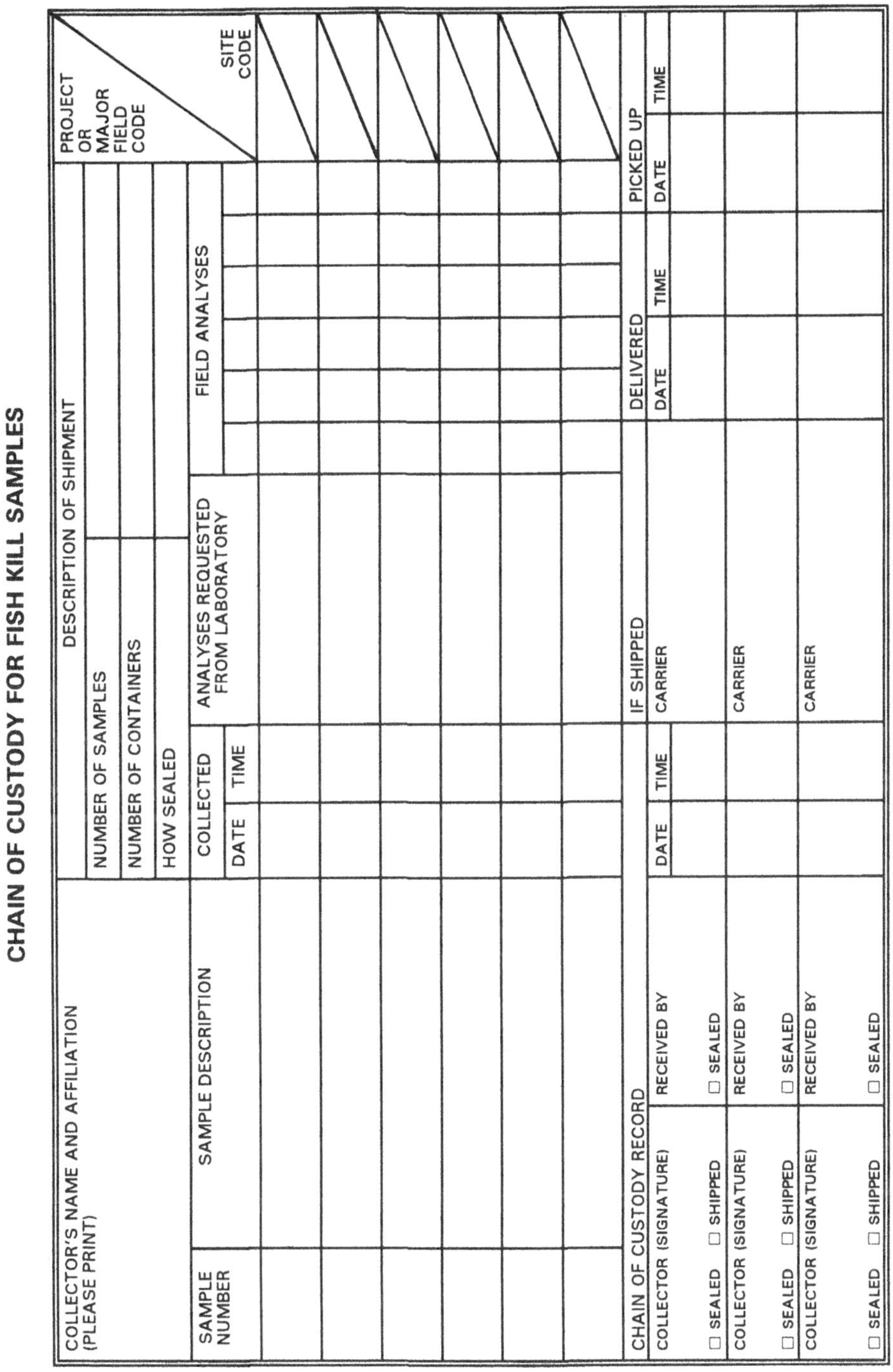

FIGURE 1.4. Example of a form for recording the chain of custody for samples taken during a fish kill investigation.

FISH KILL INVESTIGATION REPORT

Prepared and submitted by: ____________________

Date: ____________________

1. Kill reported to investigator by:

   Name ____________________

   Address ____________________

   Phone __________ Date __________ Time __________

2. Investigator reported kill to:

   Name ____________________

   Address ____________________

   Phone __________ Date __________ Time __________

   Name ____________________

   Address ____________________

   Phone __________ Date __________ Time __________

3. Body of water ____________________
4. Nearest town __________ County __________
5. Time and date of kill ____________________
6. Time and date of investigation ____________________
7. Suspected cause of kill, or describe how kill occurred ____________________

   ____

   ____

8. Alleged contamination source ____________________
9. Duration of kill ____________________
10. Extent of kill: Stream miles __________ Lake acres __________
11. Estimated total number of fish killed (from #24) __________ Value $__________
12. Species of fish killed ____________________

    ____

    ____

- 1 -

FIGURE 1.5. Example of a fish kill investigation report form.

13. Condition of fish found (check one or more):

    Dead for several days ____________ Dead for a short period ____________

    Dying and in distress ____________

    Other (write in) ____

    ____

14. Symptoms of distressed fish ____

15. Abnormal appearance of the water ____

16. Investigation personnel (give name and organization represented):

    ____

    ____

    ____

    ____

    ____

17. Other persons and organizations involved:

    Name and title | Organization and address

    ____________________ ____________________

    Remarks ____

    ____________________ ____________________

    Remarks ____

    ____________________ ____________________

    Remarks ____

18. Topographic view or map of the affected area:

- 2 -

FIGURE 1.5. Continued.

19. Chemical data:

Sample location & tag number	Sample depth	Date	Time	D.O.	pH	Temp.	Other field tests	Water samples

Chemical data prepared by: ______________________________

Title: ______________________________ Date: ______________

20. Discussion:

21. Recommendations of corrective action needed:

- 3 -

FIGURE 1.5. Continued.

22. Projected loss of fisheries:

Water body ______________________________

Number of sample areas ________ Length ________ Surface acres ________

Total length of stream affected ______________

SPECIES

____

Length	Total number	Total weight	Value per fish	Total value	Length	Total number	Total weight	Value per fish	Total value
1"					11"				
2"					12"				
3"					13"				
4"					14"				
5"					15"				
6"					16"				
7"									
8"									
9"									
10"					Subtotal				

____

Length	Total number	Total weight	Value per fish	Total value	Length	Total number	Total weight	Value per fish	Total value
1"					11"				
2"					12"				
3"					13"				
4"					14"				
5"					15"				
6"					16"				
7"									
8"									
9"									
10"					Subtotal				

Total: ______________

Prepared by: ______________________________

Title: ______________________ Date: ______________

- 4 -

FIGURE 1.5. Continued.

23. Field investigation expenses:

Agency ______________________________

Name	Hours	Salary	Miles driven	Milage allocation	Boat operation	Film	Other (itemize)
TOTALS							

Prepared by: ______________________________

Title: ______________________________ Date: ______________________________

24. Totals:

Total number of fish lost ____________

Total fishery value lost $ ____________

Total expenses $ ____________

GRAND TOTAL $ ____________

- 5 -

FIGURE 1.5. Continued.

Table 1.4. Some suggested references for a fish kill investigation library. Entries preceded by an asterisk (*) are cited in the text of Chapter 1 and are listed as well in the Reference section.

All, J. N., and M. F. Treacy, editors. 2006. Use and management of insecticides, acaricides, and transgenic crops. American Phytopathological Societ, St. Paul, Minnesota.

American Fisheries Society. 1993. Sourcebook for investigation and valuation of fish kills. American Fisheries Society, Special Publication 24 (supplement), Bethesda, Maryland.

American Society for Testing and Materials. 1996. Standard guide for fish and wildlife incident monitoring and reporting ASTM Standard Designation E 1949-96. American Society for Testing and Materials, West Conshohocken, Pennsylvania.

Ando, A. W., and M. Khanna. 2004. Natural resource damage assessment methods: lessons in simplicity from state trustees. Contemporary Economic Policy 22:504–519.

*APHA (American Public Health Association), American Water Works Association, and Water Environment Federation. 2012. Standard methods for the examination of water and wastewater, 22nd edition. APHA, Washington, D.C.

Bain, M. B., and N. J. Stevenson, editors. 1999. Aquatic habitat assessment: common methods. American Fisheries Society, Bethesda, Maryland.

Bell, G. R. 1978. Investigation of mortalities in the wild. *In* T. Bagenal, editor. Methods for the assessment of fish production in fresh waters. IBP (International Biological Programme) handbook 3, 3rd edition. Blackwell Scientific Publications, Oxford, UK.

Beyer, W. N., and J. P. Meador, editors. 2011. Environmental contaminants in biota, 2nd edition. CRC Press, Boca Raton, Florida.

Boyd, C. E. 1990. Water quality in ponds for aquaculture. University of Alabama Press, Birmingham.

Boyd, C. E. 2015. Water quality: an introduction, 2nd edition. Springer, Basel, Switzerland.

Chapman, D. 1996. Water quality assessments: a guide to the use of biota, sediments, and water in environmental monitoring, 2nd edition. E&FN Spon, New York.

Clemson Cooperative Extension. 2017. South Carolina pest management handbook. Available: www.clemson.edu/extension/agronomy/pest%20management%20handbook.html. (August 2017).

Dow AgroSciences. 2017. Product information online. Available: www.dowagro.com/en-us/products/united-states.(August 2017).

Eisler, R. 2000. Handbook of chemical risk assessment: health hazards to humans, plants, and animals. CRC Press, Boca Raton, Florida.

Geraci, J. R., and V. J. Lounsbury. 1993. Marine mammals ashore: a field guide for strandings. Texas A&M University, Sea Grant College Program, Galveston.

Hayne, D. W., R. D. Ober, L. Schaaf, and D. Scott. 1980. Barkley Lake symposium: comparison of second day pickup with numbers estimated by Pollution Committee counting guidelines—Barkley Lake study. Proceedings of the Annual Conference Southeastern Fish and Wildlife Agencies 33:738–752.

Heath, A. G. 1995. Water pollution and fish physiology, 2nd edition. CRC Press, Boca Raton, Florida.

Helfrich, L. A., and S. A. Smith. 2000. Fish kills: their causes and prevention. Virginia Cooperative Extension, Publication 420-252, Arlington.

TABLE 1.4. Continued.

*Hill, D. M. 1983. Fish kill investigation procedures. Pages 261–274 *in* L. A. Nielsen and D. L. Johnson, editors. Fisheries techniques. American Fisheries Society, Bethesda, Maryland.

Jarvinan, A. W., and G. T. Ankley. 1999. Linkage of effects to tissue residues: development of a comprehensive database for aquatic organisms exposed to inorganic and organic chemicals. Society of Environmental Toxicology and Chemistry, Pensacola, Florida.

Johnson, W. W., and M. T. Finley. 1980. Handbook of acute toxicity of chemicals to fish and aquatic invertebrates. U.S. Fish and Wildlife Service, Resource Publication 137, Washington, D.C.

Krebs, C. J. 1999. Ecological methodology, 2nd edition. Addison-Wesley, Boston, Massachusetts.

Labay, A., and D. Buzan. 1999. A comparison of fish kill counting procedures on a small narrow stream. North American Journal of Fisheries Management 19:209–214.

Meister, R. T. 1996. Farm chemicals handbook, '96. Farm Chemical Magazine, Willoughby, Ohio (out of print).

*Meyer, F. P., and L. A. Barclay, editors. 1990. Field manual for the investigation of fish kills. U.S. Fish and Wildlife Service, Resource Publication 177, Washington, D.C.

Ofiara, D. D. 2002. Natural resource damage assessments in the United States: rules and procedures for compensation from spills of hazardous substances and oil in waterways under US jurisdiction. Marine Pollution Bulletin 44:96–110.

Pohanish, R. P. 2014. Sittig's handbook of pesticides and agricultural chemicals, 2nd edition. Elsevier, Amsterdam.

Queensland Department of Environment and Heritage Protection. 1998. Fish kill reporting and investigation manual: for use in investigation of possible breaches of the Environmental Protection Action Act 1994 and Fisheries Act 1994. Queensland Department of Environment and Heritage Protection, Brisbane City, Australia.

Rand, G. M., editor. 1995. Fundamentals of aquatic toxicology, 2nd edition. Taylor & Francis, Boca Raton, Florida.

Page, L. M., H. Espinosa-Pérez, L. T. Findley, C. R. Gilbert, R. N. Lea, N. E. Mandrak, R. L. Mayden, and J. S. Nelson. 2013. Common and scientific names of fishes from the United States, Canada, and Mexico, 7th edition. American Fisheries Society, Special Publication 34, Bethesda, Maryland.

Rottman, R. W., R. Francis-Floyd, P. A. Reed, and R. Durborow. 1992. Submitting a sample for fish kill investigation. Southern Regional Aquaculture Center, SRAC Publication No. 472, Stoneville, Mississippi.

Ryon, M., J. Beauchamp, W. K. Roy, E. Schilling, B. Carrico, and R. Hinzman. 2000. Stream dispersal of dead fish and survey effectiveness in a simulated fish kill. Transactions of the American Fisheries Society 129:89–100.

Schneider, J. 1998. Fate of dead fish in a small lake. American Midland Naturalist 140:192–196.

Schwarz, C., R. Bailey, J. Irvine, and F. Dalziel. 1993. Estimating salmon spawning escapement using capture–recapture methods. Canadian Journal of Fisheries and Aquatic Sciences 50:1181–1197.

Southeastern Vegetable Extension Workers Group. 2017. Southeastern U.S. 2017 vegetable crop handbook, 18th edition. Available: www.clemson.edu/public/coastal/documents/2017SEVG.pdf. (August 2017).

TABLE 1.4. Continued.

Sheffield, S. R., J. P. Sullivan, and E. F. Hill. 2012. Chapter 6: identifying and handling contaminant-related wildlife mortality and morbidity. Pages 154–180 *in* N. Silvy. editor. The wildlife techniques manual, 7th edition. The Wildlife Society, Bethesda, Maryland.

Smith, S. B., A. P. Donahue, R. Lipkin, V. S. Blazer, C. J. Schmitt, and R. W. Goede. 2002. Illustrated field guide for assessing external anomalies in fish. U.S. Geological Survey, Biological Resources Division, USGS/BRD/ITR-2002-0007, Reston, Virginia.

Strange, E., P. D. Allen, D. Beltman, J. Lipton, and D. Mills. 2004. The habitat-based replacement cost method for assessing monetary damages for fish resource injuries. Fisheries 29:17–24.

Thomson, W. T. 2000–2002. Agricultural chemicals, volumes 1–4. Thomson Publications, Fresno, California.

Thurston, R. V., R. C. Russo, C. M. Fetterolf, Jr., T. A. Edsall, and Y. M. Barber, Jr., editors. 1979. A review of the EPA red book: quality criteria for water. American Fisheries Society, Water Quality Section, Bethesda, Maryland

U.S. Environmental Protection Agency. 1978. Fish kills caused by pollution: fifteen year summary 1961–1975. U.S. Environmental Protection Agency, Report 440/4-78-011, Washington, D.C.

U.S. Environmental Protection Agency. 1986. Quality criteria for water. U.S. Environmental Protection Agency, Report 440/5-86-001, Washington, D.C.

U.S. Environmental Protection Agency. 2017. Water quality criteria. Available: www.epa.gov/wqc. (August 2017).

*USOFR (U.S. Office of the Federal Register). 1991. Natural resource damage assessments. Code of Federal Regulations, Title 43, Part 11. U.S. Government Printing Office, Washington, D.C.

Wedemeyer, G. A., and J. W. Wood. 1974. Stress as a predisposing factor in fish diseases. U.S. Fish and Wildlife Service, Fish Disease Leaflet 38, Washington, D.C.

Zale, A. V., D. L. Parrish, and T. M. Sutton, editors. 2012. Fisheries techniques, 3rd edition. American Fisheries Society, Bethesda, Maryland.

Chapter 2
Principles of Sampling Dead Fish

A conspicuous fish kill evokes the interest of the public and the concern of resource managers, and it usually results in an investigation. In most states and provinces, the primary responsibility of the fish kill investigator is to describe the fish killed in terms of numbers, species, size and weight distributions, and monetary values. Exact identification of the cause may be determined by the fish kill investigator or by a pollution control agency if a pollutant is suspected to be involved.

A fish kill typically is unexpected and short-lived. It demands the immediate attention of a fish kill investigator who must instantly design a sound and defensible sampling scheme.

This chapter provides general background information on carrying out a field survey of visible dead fish at a fish kill, including the procedural details, definitions, fish counting methodology, and directions for calculating standard errors of estimated totals. Chapter 3 provides directions for making sampling surveys of fish kills at several sites (narrow streams, lakes, etc.) with the objective of estimating the number of visible dead fish. The two chapters provide the basic information needed to investigate a specific fish kill.

General Cautions

There is a limit on how specific these instructions can be because no guidelines are feasible for every set of field conditions. Each kill is unique and requires some adaptation of general methods. Investigators who may be forced to deviate from the methods described here should follow the principles of area sampling as closely as possible (see Area Sampling, later in this chapter). Deviations from these methods and reasons for making them should be described in the field notes and in the fish kill investigation report (see Chapter 1).

If a monetary value is to be assigned to the killed fish, Appendix A of this publication should be examined to determine whether a length or weight measurement is needed. If only lengths will be measured in the field, as often happens, length–weight conversion functions may be needed.

Generally, the procedures in Chapters 2 and 3 are explained in English measurement units, but metric units are equally appropriate. Most of the procedures describe a one-time, short-term sampling of the kill. However, problems with sampling a kill for more than 1 d are discussed.

Chapters 2 and 3 are revisions of material in American Fisheries Society Special Publication 24 (AFS 1992) and Special Publication 30 (Southwick and Loftus 2003).

These chapters do not discuss the use of aerial photography. Aerial photography has been effectively used to assess the extent of mortality events involving several different aquatic animals. With the introduction of unmanned aerial vehicles capable of collecting photographic records, it is highly likely that this technology will continue to evolve as a useful tool for documenting the spatial extent and density of dead organisms.

Estimates of losses based on countable dead fish will likely underestimate the actual total. Very seldom will the counts represent more than a modest fraction of the fish killed because the counts are based only on fish actually observed during a single window of time during a dynamic, ongoing process. Fundamental problems prevent accurate estimation of the total numbers of dead fish. Fish die at differing rates, and once dead, they float or sink at different times; for the same species and toxicant, these rates vary with water quality, temperature, and size of fish. A count of dead fish will miss many fish that are too deep in the water to be seen, are hidden by debris, have been taken by predators or scavengers, have decomposed, or are visible but overlooked (human error). These factors contribute to underestimating the numbers of fish killed.

The Nature of Fish Kills

A fish kill investigator should know and be able to use published work on fish kills and the reactions of fish to toxicants. A large body of published information on fish kills comes from the directed use of toxicants (usually rotenone) in population studies and management operations. The studies referenced here show that reaction to toxicants differs among species, among sizes and ages of fish within species, and with the physical conditions of the water.

After lethal exposure, some fish float while others sink. There is evidence that recovery of dead fish is usually incomplete. Fish kills are not instantaneous. Early studies reported differential responses to toxicants by various species and sizes of fish (Brown and Ball 1934a; Leonard 1939; Krumholz 1944).

Most population studies have used a 3-d pickup of dead fish. Davies and Shelton (1983), however, recommended picking up for 2 d, continuing for a third day if necessary. A 3-d pickup was used in two studies conducted by the Pollution Committee, one at Douglas Reservoir (Hayne et al. 1967) and the other at Barkley Lake (Axon et al. 1980). Shireman et al. (1981) used 3 d. On the other hand, Krumholz (1944) reported picking up fish for 10 d after application of rotenone, and Ball (1948) picked up for 6 d. Parker (1970) found that although most dead fish surfaced in a few days at temperatures above 60°F, it took up to 30–40 d for some fish to rise at lower temperatures.

There seems to be no general rule about the percentage of the total pickup that can be expected on a particular day, to judge from reports of 3-d or longer pickups. As calculated from the totals for the first 3 d, Krumholz (1944) picked up, on successive days, 45%, 17%, and 38% of the fish by number and 63%, 13%, and 24% by weight. Henley's (1967) data for 11 cove studies show 53%, 37%, and 10% recoveries by number on days 1–3 and corresponding weight recoveries of 64%, 25%, and 11%. In the Barkley Lake cove rotenone study (Hayne et al. 1980), recovery percentages (of the totals) varied from day to day among four subareas that had been established with block nets. The weighted mean percentages (the results if there had been no subareas) and the daily ranges for the four subareas were, for numbers of fish, 8% (8–24%) on day 1, 82% (72–88%) on day 2, and 4% (2–9%) on day 3. Respective weight recoveries were 40% (31–48%) on day 1, 49% (40–53%) on day 2, and 11% (8–13%) on day 3. Also in the Barkley Lake study, Axon et al. (1980) observed successive numerical recoveries of 40%, 49%, and 11% for marked black bass and other fish. In these studies, the higher proportions of weight than of numerical recoveries on the first day reflect earlier recoveries of the larger fish.

Several studies have further confirmed that recovery of killed fish is incomplete. In early studies, Brown and Ball (1943b) and Krumholz (1948, 1950) reported large numbers of unrecovered fish. Krumholz (1944) recovered 86% of marked fish during a 10-d pickup (71% during days 1–3 and 15% in numbers during days 4–10). In a 6-d pickup, Ball (1948) recovered 59% of the Bluegills *Lepomis macrochirus* and 45% of the Brook Trout *Salvelinus fontinalis* that had been marked 4 d before the poisoning; he also reported recoveries of 23% and 14% of fish that had been marked up to many months previously in two other studies. Henley (1967) used scuba gear in 11 Cumberland Lake coves to estimate that dead fish of all species left on the bottom after 3-d pickups amounted to 34% by number and 5% by weight of all fish killed; the percentages differed among species. Axon et al. (1980) recovered 91% of the fish marked before the 3-d Barkley Lake study. A sampling survey early the second day of the Barkley Lake experiment indicated that the dead fish available at that time were only about half the number picked up that entire day, and about 40% of the 3-d pickup (Hayne et al. 1980).

Davies and Shelton (1983) offered a table of adjustment factors for nonrecovery of 22 species and groups of species. Values range from 1.27 for Flathead Catfish *Pylodictis olivaris* to 5.00 for Bowfin *Amia calva* (work credited to Larry Aggus of National Reservoir Research Program, U.S. Fish and Wildlife Service, Fayetteville, Arkansas).

It is clear that at any fish kill, the number of dead fish available to be counted changes over at least several days. The reasons include not only those previously mentioned, but also removal of dying fish by predators and scavengers (Krumholz 1950) and decomposition of the dead fish. Additional newer references to studies that document the fate of dead fish are provided in Table 1.4.

Definitions

Definitions of common terms used in Chapters 2 and 3 are as follows:

Area sampling—A sampling approach where the area of the kill is conceptually divided into smaller unit areas that are nonoverlapping and can be listed, selected, and visited.

Expansion factor—The number that is multiplied by the number of fish collected from the sample sites in order to estimate the total number of fish in the segment. For example, an expansion factor may be calculated by dividing the area of study by the sum of the lengths of sample sites or area of sample sites. An area of study that is 1,000 yd long with three sample sites of 33.3 yd each would have an expansion factor of 10 (1,000 yd/(33.3 × 3) = 10).

Gap distance—The distance between adjacent sample sites (e.g., sample segments or sample transects). The gap distance must be the same between sample sites within a segment. To determine gap distance, divide the length of the area of study by the number of desired sample sites (at a minimum, three). For example, an area of study that is 1,000 yd and has three sample sites would have a gap distance of 333 yd (1,000 yd/3 = 333 yd).

Sample sites—The sites within the study area where fish are collected, counted, and measured.

1. When using segments, a sample site is a linear length within a segment along a stream or shoreline. Sample sites within a segment are of the same length and are evenly spaced (gap distance). A sample site can be the entire length of a segment, in which case, the expansion factor for that segment will be one.
2. When using transects, sample sites are transects located perpendicular to a survey line and are of similar width and evenly spaced (gap distance). The area of a sample site encompasses the known width and length of the transect that crosses the open water.

Sample size—Sample size includes the number of sample sites and area or distance surveyed in the study area.

Segments—The areas arranged in a linear manner along a stream or shoreline that are within the area of study. Widths of these areas may be constant, as with successive segments in a shoreline zone along a lake, or variable, as with short stretches of a narrow stream of varying width. Under ordinary conditions, segments will be of uniform length. The segments must completely cover the area of study (e.g., total stream or shoreline being surveyed) without overlapping, and their total number must be known.

Strata—A subdivision of an area of study into two or more parts that are surveyed independently of each other, and often, with varying intensity of sampling effort. Individual expansion factors may be calculated for each identified strata if needed. Multiple strata should be used in an area of study when there are different habitat types present, the distribution of fish is significantly different, or the site conditions are significantly different.

Study area—The entire length of a stream, length of shoreline of a lake, or area of a lake, where the fish kill occurred.

Survey line—The measured length along the shoreline of a lake, or running parallel to the current on a large stream, that covers the entire area of study, and from which transects are offset perpendicular.

Transects—Parallel strips of known width that do not overlap and together completely cover some wide area, such as the open water of a lake or river or a piece of marsh. Typically, transects vary in length (and, therefore, in area).

Complete Enumeration

The best way to determine the number of dead fish in an area is to count them all. Such complete enumeration (or total count), however, is rarely practical because the large numbers of fish and size of the impacted area usually vastly exceed available agency resources. Therefore, an appropriate sampling methodology must be used to obtain a representative and statistically valid sample of the impacted area.

Area Sampling

Instead of total counts, area sampling can be used to investigate fish kills. When choosing the sample size, the investigator needs to balance obtaining a statistically valid representation and the time, energy, money, labor, equipment and access available. Certain simple principles of area sampling must be followed.

1. **Sample sites.** In a fish kill, dead fish are scattered over an extended area of stream, lakeshore, or open water. The numbers of dead fish are estimated from sample sites, selected at random. All dead fish in these units are counted and measured, and the counts are expanded over the entire affected area to estimate the total number of fish in the kill.
2. **Sample sites must be chosen at random.** A sampling strategy made with the minimum of bias is the most statistically valid. To avoid the introduction of personal bias, either intentional or unintentional, the sample sites must be chosen by a completely objective method—one that, in principle, allows every possible unit a known chance of being selected for counting. Although sample sites selected by personal judgment may seem more representative of the kill, such selection harms the objectivity and the results are vulnerable to an accusation of bias. The sampling method must be as defensible as possible; therefore, sample sites must be selected at random. If sample sites cannot be selected at random, then every effort must be made to maintain objectivity in sampling. The presence of inaccessible areas does not prohibit use of random sampling in accessible areas (see Inaccessible Areas, later in this chapter).
3. **Accuracy and precision depend on the number of sample sites, total area surveyed, and number of fish counted.** A larger sample size (greater area or distance surveyed) and higher percentage of dead fish counted will result in a more accurate estimate of the total number of fish killed (Variability among Sample Sites, later in this chapter).

To use the area sampling approach, the investigator must divide the study area into sample sites that (1) do not overlap, (2) together completely cover the entire area, and (3) are of practical field size (i.e., small enough that all fish on several units can be counted and measured). Sample sites must be established and identified in such a way that sample sites can be randomly selected. Sample sites can be the same size throughout a study, but sometimes size may be varied for practicality.

When implementing the area sampling approach to count dead fish, study areas are generally divided into transects or segments; uses of these are described below. Terrestrial studies often use the two-dimensional matrix of regular or irregular unit areas, which can rarely be used in aquatic studies because of the difficulty of precisely delineating unit-area boundaries.

Transects

Transects are parallel strips of known width that do not overlap and together completely cover some wide area, such as the open water of a lake or river or a piece of marsh. Typically, transects vary in length (and, therefore, in area). The following directions are for sampling along transects for systematic sampling with a random start.

To delineate parallel transects completely covering the area of the fish kill, first establish a convenient survey line, which must be perpendicular to the direction of the transects. The length of the survey line must be known and must be long enough that the two terminal transects (one perpendicular to each end of the survey line) will just enclose the area of the kill. To minimize the variability among transect lengths, the direction of the survey line should be parallel to the long axis of a lake (Figure 2.1) or to the direction of flow in a large stream. When transects are used with a meandering stream, a single survey line parallel to the net flow may be used (Figure 2.2), or a series of survey lines may be used, one for each section of stream (Figure 2.3).

The location, direction, and length of the survey line must be known in the field before transects can be established. If good maps or digital maps are available, the survey line can be located and its length determined before starting the field investigation. If a good map does not exist, the survey line can be marked in the field by pacing with a compass/Global Positioning System (GPS), using a range finder, or other survey methods.

Once a survey line has been established, apply transect sampling in the following manner:

1. Choose a transect width so that all dead fish floating within it can be collected from a boat. A useful device for marking this width is a pole fastened across the bow and extending to one or both sides of the boat. Ropes trailing from the pole into the water mark a measured width. Collect dead fish only within this width, and use the same width consistently throughout an investigation.

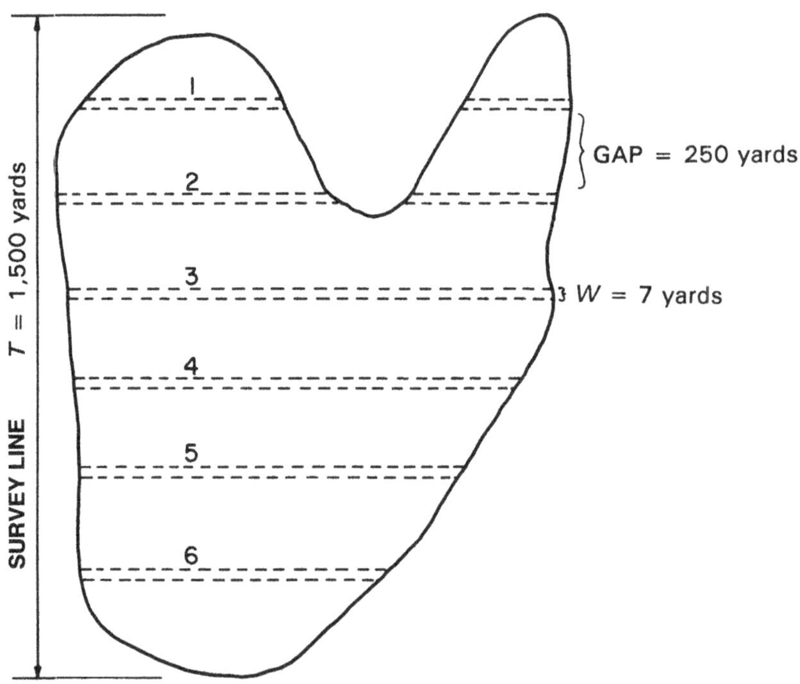

FIGURE 2.1. Survey line and transects for open water of a lake. T = survey line distance; W = transect width..

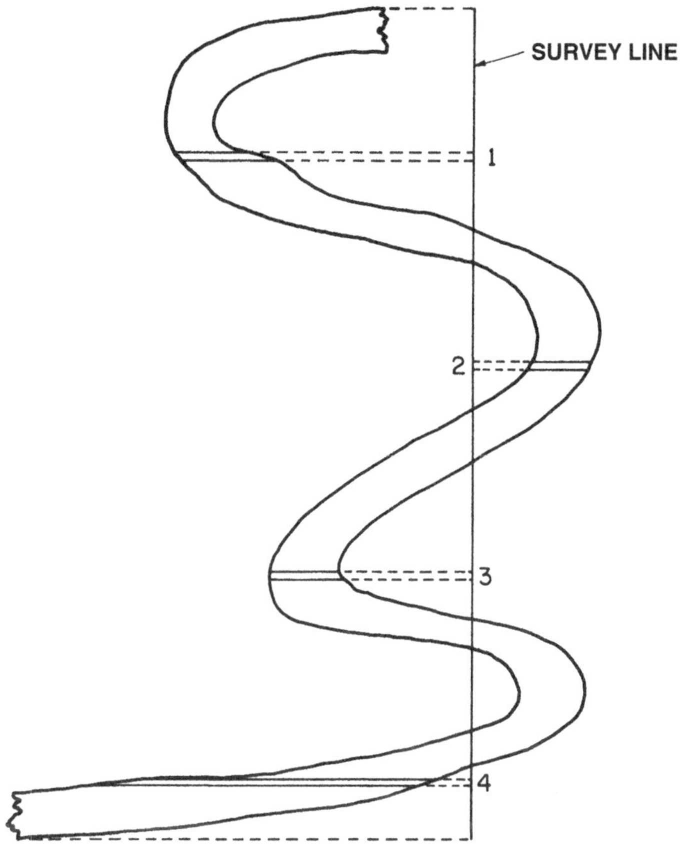

FIGURE 2.2. Survey line and transects for open water of a large stream.

2. Determine the number of sample sites (i.e., transects where fish will be collected and measured for fish counts). The greater the area surveyed and the greater number of sample sites, the more accurate and precise the result will be (Sample Size, later in this chapter). The exact number of sample sites required will depend upon the variability of the observations, as well as time and manpower; *run at least three.*
3. Calculate the gap distance (distance between adjacent sample sites) by dividing the survey line length by the number of sample sites to be surveyed.
4. Calculate the number of transects in a gap by dividing the gap distance by transect width.
5. Round both calculations to the nearest whole unit and mark the location of each sample site on the survey line. Start by selecting one sample site at random (Use of Random Number Tables, later in this chapter) from all possible transects in the first gap distance at one end of the survey line. Mark the location of this sample site on the survey line in the center of the sample site width. The next sample site will be one gap distance further along the survey line from the first sample site (center to center). The rest of the sample sites are located sequentially along the survey line, each one gap distance from its predecessor.

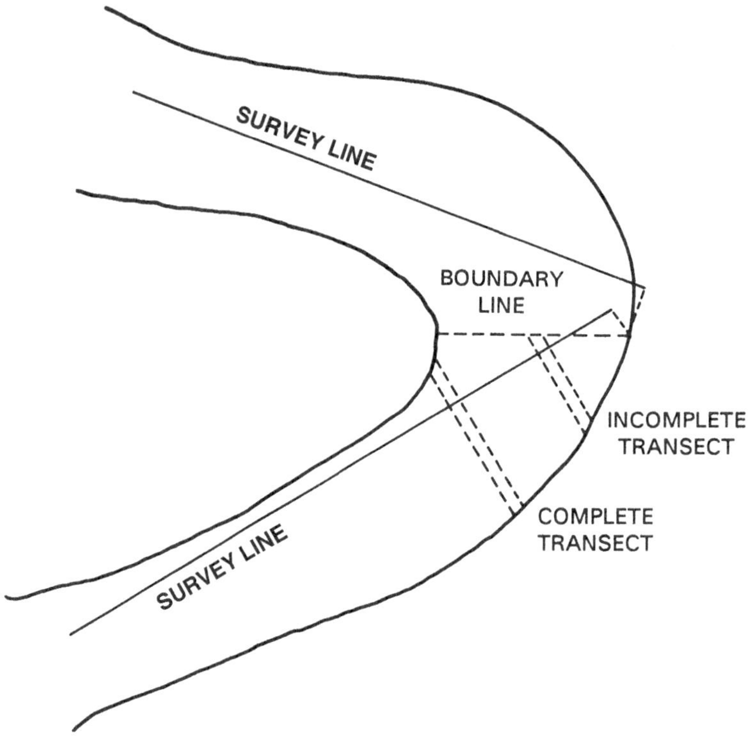

FIGURE 2.3. Survey lines and sample transects for open water of a large meandering stream when multiple survey lines are used.

Identify each location of a sample site using the shoreline or other markers with a compass, GPS, or other method perpendicular to the survey line. Mark each location on the shore and continue the transect by boat. Collect all floating dead fish within the transect width as the boat moves on the compass course from the marked shore location to the opposite shore. If there is a current, it is important to guide the boat toward a landmark that has been preselected, by compass bearing, on the opposite shore. Be aware of safety concerns while navigating a boat along each transect, but maintain the bearing as close as possible.

Transects must extend completely across the area being studied. If there is to be a separate count of dead fish along the shoreline, transects must extend from the outer edge of one shoreline zone to the outer edge of the other to avoid double-counting. If there is no shoreline stratum, transects extend from water edge to water edge.

A crew of at least two people is needed to collect dead fish on a transect. One worker must keep the boat on course at a speed slow enough that one or more other workers can use long-handled nets to collect all of the fish and place them in a tub for later examination. Record the numbers of dead fish separately for each sample site so that the total number of dead fish can be calculated.

There are two methods of using counts on transects to estimate the total number of dead fish on an open-water stratum or zone. These are illustrated in more detail with examples provided in Chapter 3 (Open-Water Counts). The appropriate method depends upon the information available.

With a minimum of information and knowing only the length of the survey line, an investigator can use Method A. All the investigator needs to know is the length of the survey line, the number and width of the sample sites, and the numbers of fish collected on those sample sites. The total number of dead fish in the study area is estimated either by multiplying

1. the average number of dead fish per sample site by the total number of transects in the area, or
2. the total number of fish on all sample sites by an expansion factor. This expansion factor is calculated either as the ratio of the total number of possible sample sites to the total number of sample sites or as the ratio of the length of the survey line to the combined total width of the sample sites.

Both calculations will produce the same result.

If there is adequate information for determining both the area of the body of water and the length of each sample site, the investigator should use Method B because it increases the precision and accuracy for the same amount of fieldwork. The investigator must know the area of the body of water, the width and total length of all sample sites, and the total number of fish on all sample sites. The total number of dead fish floating in the open-water zone is estimated as the product of the area and the density of dead fish. Density (fish per unit area) is the total number of fish from all sample sites divided by the total area of all sample sites (transect width multiplied by the total length of all sample sites). If Method B gives the better precision and accuracy with the same number of transects, why should Method B not always be used? First, for Method B to excel, the transect lengths must differ and there must be a good correlation between transect length and number of fish. These conditions are likely with most irregularly shaped water bodies but not with rivers of uniform width. Second, unless the requisite measurements (e.g., length, area) can be gained inexpensively, the added cost of this information can so reduce the affordable number of sample sites that there is no gain in precision. However, with new technology and information available, this is becoming less of an issue.

Transects also can be used to derive a crude estimate of the area of the body of water covered *if* the length of the expanse of water covered by each sample site is known. The area of water covered by each transect is calculated as the length over water multiplied by the transect width. The mean area covered per sample site multiplied by the total number of transects estimates the total area. Note, any attempt to use the area calculated in this way to implement Method B reduces mathematically to Method A.

Segments

Segments, as used here, are areas arranged in a linear manner. Widths of these areas may be constant, as with successive segments in a shoreline zone along a lake, or variable, as with short stretches of a narrow stream of varying width. Under ordinary conditions, sample sites will be of uniform length, although segments of varied length may be more practical under certain conditions. The segments must completely cover the area of study (e.g., total stream or shoreline being surveyed) without overlapping, and their total number must be known. Methods of selecting a sample from segments of uniform length will be described first.

To establish segments that are uniform in length, first decide what the segment length will be. The objective is to have segments that are long enough to contain a useful number of fish yet not so long that it will be impractical to count all the fish on several sample sites. Aim for 50–200 fish per segment in the more dense areas of the kill. Next, measure on a good map or digital mapping tool the total length of the shoreline or stream to be surveyed, or, in the field, measure this length with a tape or other method. Next, decide the number of sample sites on which fish will be counted and measured. Include at least three sample sites, but use more if possible (Sample Size, later in this chapter).

Locate the sample sites as a systematic sample with a random start (later in this chapter). Determine the gap distance (the length of the study area divided by the number of sample sites to be used). Determine the number of nonoverlapping segments that can be contained within a gap distance (round to the nearest whole unit). Select at random one sample site in the first gap distance, starting at one end of the area to be surveyed (Use of Random Number Tables, later in this chapter). Measure (or count) one gap distance to locate the second sample site and then subsequent sample sites. Keep separate records of the number and size of dead fish on each sample site so that a standard error (later in this chapter) can be calculated for the total number estimate.

If segments are of equal length, the total number of dead fish in the area of the kill is estimated one of two ways:

1. Multiplying the average number of fish per sample site by the total number of segments in the area of the kill.
2. Multiplying the total number of fish in all sample sites by an expansion factor. The expansion factor can be calculated either as the total number of segments divided by the total number of sample sites or as the total length of all segments in the area of the kill divided by the total length of all the sample sites.

Under some field conditions, the length of the segments can be varied to improve sampling effectiveness. For example, dead fish ordinarily are scattered irregularly along a lakeshore. If the segments are lengthened where fish are scarce and shortened where fish are numerous, with the objective of having roughly the same number of fish per segment (e.g., 200), the variation in numbers per segment will be reduced and precision of the estimate will be increased for the same number of sample sites. This procedure requires making a preliminary trip along the shore (or stream), marking the segment boundaries with temporary stakes, and counting the total number of segments. Then, decide the number of sample sites where dead fish will be counted, and divide the total number of segments by this number to determine gap distance, stated in number of segments (not in distance). Select at random one sample site from the first gap distance; then, count the gap distance (in number of segments) along the shore to locate the second sample site and then subsequent sample sites.

If segments are of unequal length, estimate the total number of fish killed either as the average number of fish per sample site multiplied by the total number of segments or as the total number of fish on all sample sites multiplied by an expansion factor. The expansion factor is the total number of segments divided by the total number of sample sites. Do not calculate the expansion factor on the basis of segment length when segments are of unequal length.

The method of making segment lengths inversely proportional to densities of dead fish can reduce the standard error of the estimate for a given sample size (number of sample sites), but the preliminary work of setting up the sampling frame will reduce the effort available for counting dead fish; thus, it will reduce the sample size. Therefore, a judicious division of effort must be made to maximize precision overall.

Parts of a fish kill may be poorly accessible, as when a narrow stream can be reached only where it is crossed by a road and other access is prevented by vegetation, fences, or distance. Then, if area sampling with segments is being used, it may be useful to stratify and sample parts (strata) of a fish kill less intensively than others or some parts not at all. Inaccessible areas of a fish kill do not rule out the use of random sampling, however (Inaccessible Areas, later in this chapter).

Systematic Sampling with Random Start

Sampling methods described here use systematic sampling with a random start rather than completely random sampling. With a systematic sample, the sample sites (here, transects and segments) are arranged in a line and every nth unit is included in the sample. A random start makes the systematic sample an unbiased single-item random sample composed of sample sites that, after the first, are not selected at random. With simple random sampling, in contrast, the units are listed in some manner, each sample site is selected at random, and each unit has the same probability of inclusion in the sample.

With a systematic sample, the sample sites are spaced evenly throughout the area being sampled; with simple random sampling, the sample sites are scattered in a random, often highly uneven pattern. With a systematic sample drawn from a kill area, where the dead fish usually have a clumped distribution, the variability among sample sites will generally be greater than the variability among units of a simple random sample in the same area. The systematic sample shows greater internal variability because it never includes nearby units, which tend to resemble each other more closely than do distant units.

Systematic sampling with a random start has several advantages over simple random sampling in the present application. In the field, a systematic sample made up of several equally spaced sample sites is more practical, it is easier to understand, and the location of the sample sites is more economical of effort than with a simple random sample. To many field investigators, the even distribution of sample sites makes the systematic sample seem intuitively to be more representative than the sometimes very uneven distribution of sample sites selected completely at random. When a distribution is clumped, the variability among the *means* of several systematic samples probably will be less than the variability among means of several simple random samples with the same number of sample sites. Because the variability among means of samples determines the standard error, the systematic sample is more precise under these conditions.

Systematic sampling is more precise because variability (measured as the sum of squared differences from a mean) is additive. Total variability among all the sample sites (transects or segments) making up an area may be divided into two additive portions: first, the variability within samples; second, the variability among sample means. Because total variability is fixed, when variability within samples is greater (as with systematic sampling), the variability between sample means is less.

The main difficulty with a systematic sample is that, as noted above, it does not provide reliable evidence of its own precision, even though with a random start it is unbiased, and with a clumped distribution it is more precise than a simple random sample. Nevertheless, to provide a rough idea of precision, considered to be adequate for most field counts of dead fish, methods of approximating a standard error are provided. These standard errors should not be relied upon for constructing confidence limits or testing whether apparent differences are reliable.

When a more reliable estimation of precision is required, an investigator should consider use of multiple (replicated) systematic samples, each with a random start, rather than simple random samples. This will retain the greater field efficiency and greater precision of systematic sampling and allow direct observation of the variability between samples that is needed to estimate a valid standard error.

Estimating Total Number of Dead Fish

Estimation of the total number of visible dead fish is the objective of area sampling and fish counting. This section specifies an estimator (a mathematical statement used in making an estimate) appropriate for each of the three versions of area sampling described in Chapters 2 and 3. Estimators differ in detail among the three sampling versions. Each estimator is stated first as a more easily understood procedure and then repeated as an arithmetically identical expansion factor for the number of fish counted on all sampling units. Use of expansion factors is attractive when calculation is by hand because of the time saved. One calculation (e.g., a mean) is saved for each category for which an estimated total is desired, and categories (species, size, etc.) may be numerous. Note, however, that even though an estimate can be made from the total fish collected, calculation of a standard error requires records for each sample site.

In the first unmodified version of area sampling, sampling units are of equal size (segment length or transect width; no attention is given to the other dimensions). To estimate the total number of dead fish, the average number of fish per sample site (including all zero counts) is multiplied by the total number of units (segments, transects) in the kill area. Or, the total number of dead fish counted on all the sample sites can be multiplied by the expansion factor. In this case, an expansion factor can be calculated in two ways. One way is to divide the total number of units in the kill area by the number of sample sites. The other way is to divide the total linear expanse being surveyed by the total for the appropriate dimension of the sample sites: the survey line length divided by the summed width of sample sites, or the stream or shoreline length divided by the summed length of sample sites. For example, the second type of expansion factor, for a stream survey based on segments measured in English units, is

$$\frac{\left(1{,}760\ \frac{\text{yd}}{\text{mi}}\right)\times\left(\text{total miles of all segments in the kill}\right)}{\text{summed length of sample sites, in yards}}.$$

A corresponding metric expansion factor is

$$\frac{\left(1{,}000\ \text{m/km}\right)\times\left(\text{total kilometers of all segments}\right)}{\text{summed lengths of sample sites, in meters}}.$$

In the second version of area sampling, segment length is deliberately varied to reduce the variability among segment counts of dead fish and, consequently, to increase precision (see Segments, earlier in this chapter). The total number of dead fish is estimated as the average number of dead fish per sample site multiplied by the total number of segments in the kill area. Or, the total number of dead fish on all segments can be multiplied by an expansion factor. In this case, the expansion factor is the total number of segments in the kill area divided by the number of sample sites. Note that here an expansion factor based on segment length introduces improper weighting and a biased estimate.

In the third version of area sampling, a ratio estimate is used with transects. It may be possible to increase the precision of the estimate of the total number of fish killed if the area of the kill and of each sample site can be measured (Method B, assuming that a good map is available; Transects, earlier in this chapter). The total number of dead fish is then estimated as the area of the fish kill multiplied by mean density of dead fish on the sample sites (R), where

$$R=\frac{\text{total number of dead fish on all sample sites}}{\text{total area of all sample sites}}.$$

The total number of dead fish may be estimated by the arithmetically identical use of an expansion factor: the total area of the kill divided by the total area of all the sample sites. The total number of dead fish counted on all sample sites is then multiplied by this expansion factor. An approximate standard error of this estimate may be calculated by a method appropriate for ratio estimates (Standard Error with Ratio Estimate, later in this chapter). Do not use here the approximation that employs the mean square successive difference.

Precision of the estimate can be increased by using the ratio estimate only if the transects vary in size and if there is an appreciable correlation between the count of dead fish and the area of a transect (as should be expected with counts of fish on transects of differing areas).

Stratified Random Sampling

Stratification in sampling refers to the subdivision of a study area into two or more parts called "strata." Each stratum is surveyed independently, often with varying intensity of sampling effort.

A separate estimate is made of the total number of dead fish within each stratum, based on the sample sites within that stratum. The total number of dead fish is the simple sum of stratum estimates. If a standard error is desired for this overall estimate, a standard error must first be estimated for each of the stratum totals. The stratum standard errors are squared, the squared standard errors are added, and the square root of the sum is the standard error for the total number of fish. Stratified sampling has several advantages.

1. Stratification can result in greater overall precision (smaller standard error) than use of simple random sampling over the entire universe of study. With stratified sampling, the overall standard error is based upon a pooling of variability within the several strata. With simple random sampling over the entire universe, there is an added component of variability—the variability among stratum averages—that disappears from the sampling error when stratification is used. Thus, if strata differ, stratification can increase precision.

   Stratification can promote precision in a second way, by allowing the total number of sample sites to be distributed among the strata in any proportion required. To minimize the overall standard error (given equal cost for all sample sites), the allocation of units to any stratum should be proportional to the product of stratum size (number of unit areas) and the standard deviation of counts for unit areas within that stratum. Standard deviations are seldom known for a fish kill, but a useful rule of thumb for field use is to allocate sampling effort among strata in proportion to the total number of dead fish expected in each. For planning purposes, the total number of dead fish and their location can be crudely approximated by inspection before the survey. If the investigator really cannot approximate the totals in each stratum or if the strata have been designed to have about equal numbers of dead fish, then sampling effort should be allocated equally among the strata.
2. Stratification allows part of the total survey to be made by one field method and part by another. For example, with a lake or wide stream, fish can be counted in sample sites along the shore and separately in sample sites in the open water.
3. Separate estimates of the total number of dead fish can be made for different strata of the fish kill. Although information for separate strata may be interesting, these single-stratum estimates usually have less precision than an overall estimate for the fish kill because each stratum has a smaller sample size than the entire area of the kill. It is possible, however, to allocate sampling effort more heavily to some important stratum at the expense of less precision for the others and perhaps for the total.
4. If part of the area of the fish kill is inaccessible, stratification allows this part to be set aside as a separate stratum, not sampled and therefore unknown (Figure 2.4). The rest of the kill is surveyed in a standard manner. A statistically defensible estimate can be made only for that part of the kill that can be surveyed. For the inaccessible areas, the investigator may offer a reasonable personal estimate of dead fish numbers, based upon other observations and professional judgment, although unsupported by actual data. Under these circumstances, this is the best available information (Narrow

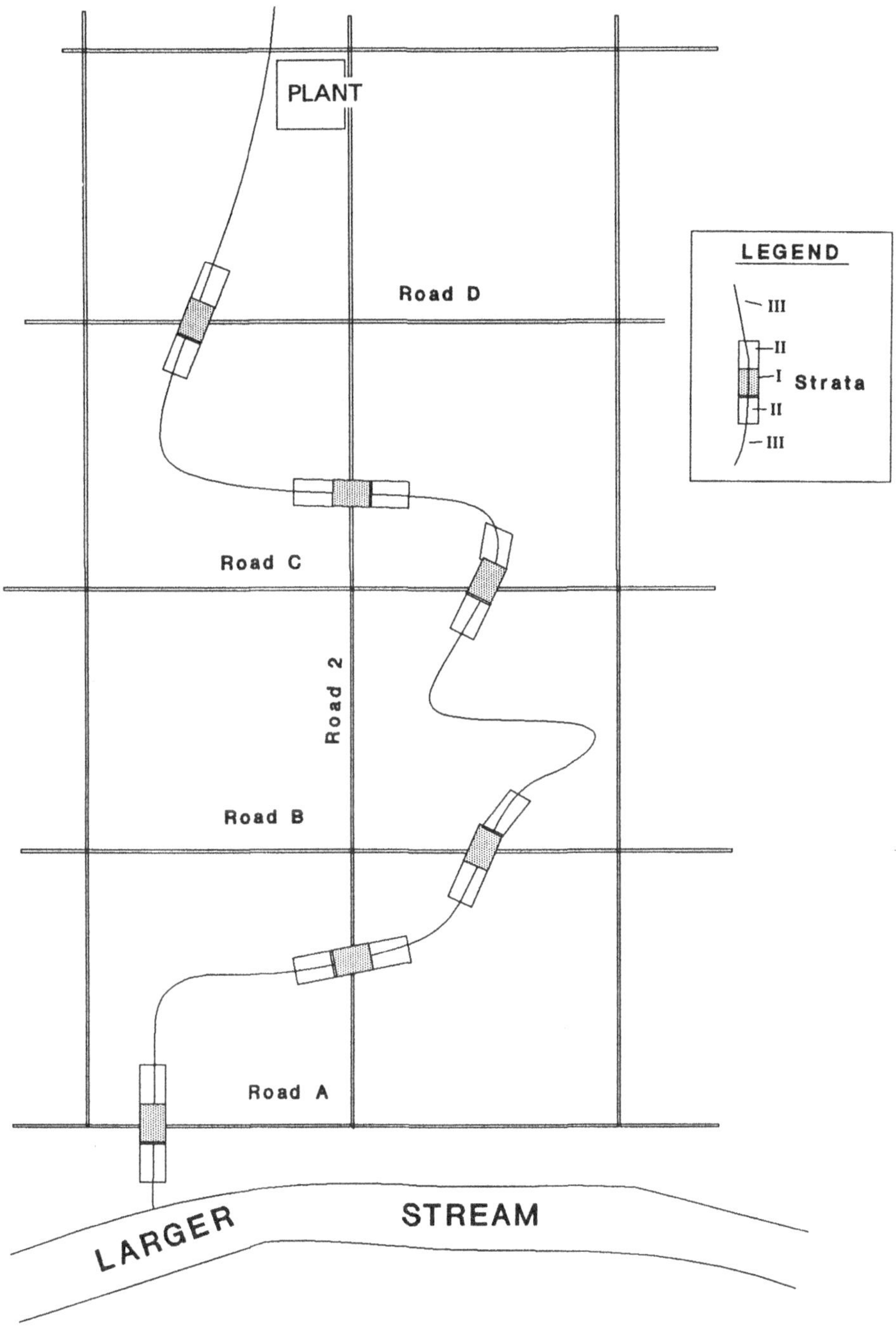

FIGURE 2.4. Sampling strate for an incompletely accessible stream. Strata III represents the inaccessible portion of the study area.

Streams, Incompletely Accessible, and Inaccessible Areas, in Chapter 3). Strafication combined with judgment also may have to be used when strata overlap to some unknown degree (Multiple-Day Counts, in Chapter 3).

5. Stratification ensures that each stratum receives a designated share of the sampling effort and avoids the chance concentration of samples that is possible with simple random sampling.

Stratified random sampling is not always useful. If the strata are poorly delineated or if sampling effort is not allocated well, the precision may be worse with stratified random sampling than with simple random sampling (but the estimation remains unbiased). Data management is more complex with stratified random sampling because each stratum has an independent survey with a quota of sampling effort. If the objective is only to estimate the total number of dead fish in the kill, then a stratum can have as few sample sites as one. If a standard error is to be approximated, then each stratum must have at least two sample sites. If there are several strata and a low budget for sampling effort, the requirement of two sample sites per stratum can result in an oversampling of the unimportant strata, a consequent undersampling of the important ones, and a reduction or loss of precision.

Inaccessible Areas

Parts of a fish kill area may not be accessible to the investigator charged with appraising the kill. Although partial inaccessibility prevents completion of a sampling survey for the entire area, it should not discourage use of random sampling in the accessible areas.

The suggestion made here is that the kill area be divided into two or more strata: one stratum where no sampling can be done and the other(s) where random sampling can be used. From a practical point of view, some reliable information is better than none. Further, experience in the partial survey may provide the investigator with support for a professional judgment about the inaccessible area.

In undeveloped areas, passage of the investigator on foot or in a boat may be impeded by heavy vegetation, fallen timber and logjams, or wetlands. In developed areas, fences, trespass restrictions, or human activity may prevent access. Remoteness itself may render an area essentially inaccessible if the cost of reaching it is too high. Anything that increases the cost per sample site inhibits sampling.

Stratification can be used not only to set aside a single unmeasurable stratum, but also to accommodate varying costs per sample site: stratum sample size may be varied according to cost per unit (as well as other factors). A similar design is suggested in Chapter 3 (Narrow Streams, Incompletely Accessible), whereby road crossings of a stream (Stratum I) are sampled intensively, the nearby accessible stream segments (Stratum II) less intensively, and the inaccessible portions of the stream (Stratum III) not at all.

How useful is a report based partly on scientific sampling and partly on professional judgment? In the abstract, a purist might fault the survey's quality. In reality, however, a manager who needs information to plan rehabilitation of the fish population likely will respect the judgment of the pollution investigator on the scene and will understand that a more complete report would require a more expensive field investigation. If the results of the survey are to support negotiations with a polluter over damages, the administrator can explain frankly that results of greater statistical quality could be obtained only at higher investigative cost, a cost that would be passed on to the polluter.

Drifting Fish in a Stream

Fish drifting downstream can cause potentially confusing problems. This movement is a flow of fish from one spot to another. The sampling surveys described in this and the following chapter produce counts of the fish visible at the moment—that is, instantaneous counts. Drifting fish can be included only if they are counted the same way: instantaneously. Any drifting fish should be counted at the moment the observer passes the fish.

The problem of correcting counts of drifting fish in the special case of narrow streams is considered in Chapter 3. The suggested procedures combine differential counts of those fish the observer passes, those fish that pass the observer, and those fish that pass a fixed point on the stream within a certain time. In principle, these methods could be applied with any transect that proceeds either directly up or directly down a wide stream (but use of such a transect would only rarely be appropriate).

In general, counts of drifting fish require correction because the water movement alters the effective area of the sample site, not because the fish are moving. The effective area of a sample site (hereafter, "transect") is the area of surface water, within transect bounds, observed by the investigator who covers the transect. The number of dead fish found equals the effective area multiplied by the density of fish (number per unit area) plus a sampling error. Thus, the essence of the problem is the effect of water movement on effective area.

With no current, the effective area of a transect is the transect length, L, multiplied by the transect width, W, or LW. If the water is flowing, however, the effective area must be adjusted for the component (vector) of flow parallel to the transect direction. To make this adjustment, define v as water velocity (distance per unit time), t as time for the observer to transit the transect, and θ as the angle at which the transect crosses the stream. The angle θ is the lower left of the four angles at the intersection of transect and streamflow when flow is represented as moving down the page and the observer is moving from left to right. Angle θ can take values between 0°, when the observer is moving directly upstream, and 180°, when the observer is moving directly downstream.

Under these conditions, the effective area covered within the transect is

$$LW + Wtv \text{ cosine } (\theta), \text{ or } W\,(L + tv \text{ cosine } \theta).$$

Cosine (θ) is positive for θ = 0–90°, zero for θ = 90° exactly, and negative for θ = 90–180°. Thus transects headed upstream have an effective area greater than transect area LW, and those headed downstream have an effective area less than LW. When a transect is perpendicular to the current (θ = 90°), no correction is needed (cosine 90° = 0). The observer may count only the fish he or she passes within the bounds of the transect.

Where a transect cuts across flowing water at an angle other than θ = 90°, there does not seem to be any easy and practical correction procedure. Measurements of water velocity and angle of crossing, while not impossible, seem impractical under most field survey conditions. Further problems arise when θ is between 90° and 180°. Under these conditions, it may be difficult to discriminate between drifting fish that the observer passes and drifting fish that pass the observer. When the water velocity is high enough, the effective area of the transect becomes negative. The observer passes no fish—all fish pass the observer—so there is no count to adjust. This is equivalent to an observer in a narrow stream moving downstream at a rate less than the water velocity, in effect backing up relative to the water. Differential counts of fish passed by and passing by the observer allow an estimate for narrow streams (Chapter 3), but for the wide-stream case, no solution is immediately apparent. The mathematical relationship described above may be useful, however, in judging the direction and relative magnitude of the count error.

Precision

The precision, or statistical reliability, of an estimate is measured by the standard error or by the square of the standard error (the variance of the estimate). The smaller the standard error, the greater the precision attributed to the estimate. The standard error is a number determined by both the variability of the data and the sample size. Precision can also be expressed as a proportional standard error (standard error/the estimate).

Explaining how the standard error provides information about precision requires a brief excursion into theory. Precision of the estimate refers to how closely repeated estimates are grouped about the true value when these estimates are made with the same methods applied repeatedly to the same population of values. Neither the true value nor the distribution of these hypothetical repeated sample estimates about the true value can be known. The observed standard error of the estimate can, however, be accepted as an estimate of the standard deviation of the distribution of repeated sample estimates about the true value—that is, as a measure of the closeness of the distribution. To go further and make a quantitative statement in probability about the distribution of estimates (i.e., confidence limits) requires further assumptions about the distribution (e.g., statistical normality of the distribution) that may not be supported by proof. Nevertheless, even without this missing information, calculated standard error of the estimate can be used as an approximate relative measure of precision.

The standard error of an estimated number of dead fish is determined by the variability among the sample counts and the number of sample sites counted. Precision can be increased by reducing the variability among sample sites or by increasing the sample size (number of sample sites). Because variability among sample sites is usually accepted as a given (but see Variability among Sample Sites, later in this chapter), increasing the precision usually means increasing the sample size and, therefore, increasing the cost and time needed to conduct the fish kill investigation.

There are simple statistical formulas for determining the required number of sample sites. To use these formulas, however, adequate precision must be defined and the variability (variance) of sample counts of dead fish must be known. Only modest precision should be attempted for estimates of total numbers and total values of dead fish after a kill. Estimates for less inclusive categories should be accepted with less precision. One criterion of modest precision is a proportional standard error of 0.15 (standard error divided by the estimate of the total number of dead fish in the kill). Even this criterion may require too much sampling effort for many fish kills because of the inherent large variability in sample counts of dead fish.

The Barkley Lake study (Hayne et al. 1980) showed how difficult it is to attain good precision in counts of dead fish. A bay was divided into four areas, three of which were sampled with (in all) 72 shore segments and 44 transects, a much more intensive effort than any agency could support for most fish kills. Even so, proportional standard errors were less than 0.15 only for a few combinations of areas, species, and sample size.

Sample Size

In planning a survey, it would be desirable to know in advance how many sample sites should be counted for a given precision. In theory, the precision required for the estimate is set first and then the necessary sample size is decided. In reality, the time, staff, and budget for sampling a fish kill usually has already been set, limiting the sample size. The question then becomes whether the precision attainable will be useful.

Counts of dead fish may be extremely variable, and the sample size required for a precision ordinarily considered scientifically respectable may be prohibitively expensive. This is especially true when the numbers of fish in size-classes of an uncommon species are estimated. Thus, good precision for the estimates means a high cost to the investigating agency and, when these costs are passed along, to the responsible party that caused the kill.

It is useful to speculate about the sample size required with a random distribution of dead fish, even though an apparently random distribution of fish is rarely encountered after a kill and sample sizes adequate for a random distribution are almost surely too small for the clumped distributions of dead fish typically encountered. With a random (Poisson) distribution, the expected proportional standard error is the reciprocal of the square root of total number counted, regardless of the number of sample sites counted (although, of course, the more areas counted, the more fish will be seen). If dead fish are randomly distributed, at least 45 fish must be counted to yield an expected proportional standard error as low as 0.15. Therefore, after a fish kill, a sample with a total of less than 45 fish usually will not meet the criterion of 0.15. Because the distribution of fish in sample sites after a kill is usually more variable than it would be if the distribution were random, the criterion of 0.15 will be exceeded.

The appropriate balance between cost and precision will vary with the kill. When the purpose of counting the fish is to determine monetary values of those killed, the survey estimates must, above all, be defensible, and this means unbiased. Although the level of precision accepted should be an administrative judgment, it seems reasonable that precision be only modest for the overall estimates and poor for other less inclusive categories. As a practical approach, a minimum of three rather large samples is suggested at several points in Chapters 2 and 3. This sample size usually will not provide good precision and may not be adequate for many needs.

Variability among Sample Sites

The high variability among counts on sample sites reflects the variability in the spatial distribution of dead fish. Unpredictable actions of currents and winds cause clumped distributions of fish that are more variable than random distributions. A distribution more regular than random would have less spatial variability, but a regular distribution is not to be expected after a kill.

If variability among the sample sites can be reduced, the precision of an estimate will be increased without an increase in sample size. Three ways of doing this are considered. Overall variability can be reduced by stratification. In the special case of segments, variability can be reduced by making the length of a segment inversely proportional to the apparent density of dead fish. Last, increasing the area of sample sites of the same size will reduce the relative variability in sample counts of dead fish.

Stratification (Stratified Random Sampling, earlier in this chapter) can increase precision under the right conditions. Stratification does not reduce the inherent variability of counts among sample sites as much as it allows the identification and exclusion of the variability that is due to differences among strata. Each new stratum, however, requires a quota of sample sites. Use of this procedure may be limited by the total number of sample sites that can be counted.

When segments are used, it is possible under some conditions to vary the length of the individual segments with the objective of reducing the variation in number of dead fish per segment (Segments, earlier in this chapter). For example, if dead fish are scattered along a lakeshore, the investigator can make a preliminary trip and mark segments that contain about 200 fish each. Some of these segments will be much larger than others, but the variability among sample counts will be reduced to the degree that the investigator can approximate 200 fish per segment on the run, so to speak. The preliminary work will consume some of the available field effort and thus reduce the number of segments that can be counted.

A sample made up of larger sample sites tends to have a proportional standard error of the total that is less than that of a sample with the same number of smaller sample sites. Some investigators have asked whether this tendency can be used to increase sampling efficiency. A partial answer can be provided by statistical theory. If dead fish were distributed at random, then the relative variance (square of the ratio: standard deviation/mean) among sample sites would decrease as a function of the reciprocals of the areas of the sample sites. If the cost of recording the data from sample sites is exactly proportional to the area of units and if the sample size is determined by the total cost of data recording, then the trade-off between using fewer larger units and more smaller units would be even. Dead fish, however, are usually clumped, rather than distributed at random. Therefore, the relative variance among larger sample sites, although less than that among smaller sample sites, does not decrease as fast as it would if the dead fish were distributed at random. This fact favors the use of a greater number of smaller units over the selection of fewer larger units. There is little to be gained from manipulating the area of the unit to reduce variability when units are the same size.

Approximate Standard Error

Systematic sampling with a random start, as used here, is really a single-unit random sample. Although this sampling method provides an unbiased estimate, it does not allow calculation of a valid standard error. However, an approximation to the standard error may be calculated under the assumption that the multiple parts of the single-unit random sample behave as though they were random samples. This approximation probably overestimates the true standard error. Recognized for what it is, however, this form of the standard error can give realistic, if approximate, guidance about precision.

Directions are presented for calculating an approximate standard error for each of the two sampling designs that have been presented. The more common design is based on numbers of fish per sample site. The second design is based on density of fish per area of sample sites and uses a ratio estimate.

Standard Error with Mean Square Successive Difference

With systematic sampling counts of dead fish on transects and segments, it is appropriate to estimate the approximate standard error by the method of mean square successive difference because successive counts often exhibit trends and are correlated. In a stream, counts may be larger upstream and decline downstream; in a lake, fish are generally more dense along one shore compared to the other shores; and nearby counts tend to be correlated even though there may not be a single trend in numbers.

If the total number of fish in the kill is estimated as the mean number of fish per sample site multiplied by the total number of units in the kill area, the approximate standard error (SE) of the sample mean is calculated by

$$\text{SE of sample mean} = \sqrt{\left(\sum d^2\right)/2N(N-1)},$$

where

d = the difference between successive counts
N = the number of sample sites in the sample series

The standard error of the estimated total number of dead fish is the standard error of the sample mean multiplied by the total number of units from which the series of sample sites was drawn.

If the total number of fish in the kill is estimated as the product of the total number of fish counted on the sample units and an appropriate expansion factor, the approximate standard error of the total number counted is required. This takes this form

$$\text{SE of total sample counts} = \sqrt{N\left(\sum d^2\right)/2(N-1)},$$

where d and N are defined as above. The standard error of the estimated total number of dead fish is the standard error of the sample count multiplied by the expansion factor used in expanding the sample count to the total number of fish killed. For both of these expressions, the equivalent number of degrees of freedom is less than $N - 1$ and probably greater than $(N/2) - 1$.

Example: On the lake (Figure 2.1), 300 dead fish were collected on six transects; an expansion factor of 35.7 led to an estimated total of 10,700 dead fish (Method A, assuming that a good map is not available). The numbers collected on each transect allow the standard error of this total to be approximated as ±2,440 by way of mean square successive difference.

Segment	Count	Difference (d)
1	14	
2	67	–53
3	107	–40
4	55	52
5	29	26
6	28	1
Total	300	

$$\sum d^2 = 7{,}790.$$

$$\text{SE}(\text{total sample count}) = \sqrt{\frac{6(7{,}790)}{2\times 5}} = 68.37 \text{ fish}.$$

$$\text{Expansion factor} = \frac{\text{total number of segments}}{\text{number of sample segments}} = 35.7.$$

When both the total count (300 fish) and the standard error of the count (68.37 fish) are multiplied by the expansion factor, the total number of fish in the kill ± standard error, rounded, is 10,700 ± 2,440 fish (proportional standard error, 0.228). If a random sample is assumed, the standard error of the total would be ±2,970 fish (proportional standard error, 0.277).

Standard Error with Ratio Estimate

When density of dead fish per unit area can be determined with a ratio estimate and expanded to a known total area of fish kill, the approximation of a standard error proceeds in a different manner. For this method, the area of the fish kill and the area of each sample site must be known. Further, for there to be any gain in precision over simple random sampling, the sample sites must differ in area and there must be a correlation between the area of the sample site and the number of fish counted on it. The density per unit area, R, is the ratio estimate, which is calculated by the formula

$$R = \frac{\text{sum of fish counted in all sample sites}}{\text{sum of the areas of all sample sites}}.$$

The total number of dead fish is estimated as the product of R and the area of the kill. The approximate standard error of this estimated total is based upon the summed, squared differences between the number observed on a sample site and the number predicted as the product of R and the area of that sample site. The approximate standard error of the estimated total takes this form:

$$\text{SE of estimated total} = U\sqrt{\sum (D)^2 / N(N-1)},$$

where U = the total number of units in the entire area of the kill; D = deviation, or difference, between the count observed and that predicted as $R \times$ (area of sample site); and N = number of sample sites in the sample series.

This method of calculating a standard error for a kill estimate derived by a ratio estimate is approximate under the best of conditions and is vulnerable to the effects of the small sample sizes.

Example: An example of making an estimate by ratio estimation is presented in Chapter 3 (Method B, assuming that a good map is available; Figure 2.1). A lake of area measured as 247.9 acres (1,200,000 yd²) was sampled with six transects, each 7 yd wide and of varying, measured length. Altogether, 300 fish were counted on the 31,850 yd² of the sample, giving estimates of 0.0094192 fish/yd² and a total of 11,300 fish (rounded). An approximate standard error of ±2,120 fish is estimated below by the method just described.

The data for this fish kill are as follows:

Transect			Fish count		Deviation
Number	Length (yd)	Area (yd²)	Observed number	Predicted number	(observed – predicted)
1	400	2,800	14	26.4	−12.4
2	800	5,600	67	52.8	14.2
3	1,000	7,000	107	65.9	41.1
4	800	6,300	55	59.3	−4.3
5	850	5,950	29	56.0	−27.0
6	600	4,200	28	39.6	−11.6
Total	4,550	31,850	300	300.0	00.0

$$R = \frac{\text{Sum of fish counted in all sample sites}}{\text{Sum of the areas of all sample sites}} = \frac{300 \text{ fish}}{31{,}850 \text{ yd}^2}$$
$$= 0.0094192 \text{ fish/yd}^2.$$

$$\sum(\text{Deviation})^2 = \sum D^2 = 2.926.7.$$

$$\text{Total number of transects} = \text{survey line length/transect width}$$
$$= (1{,}500 \text{ yd}) / (7 \text{ yd/transect})$$
$$= 214.3 \text{ transects.}$$

$$\text{SE of total number} = 214.3\sqrt{\frac{2{,}926.7}{6 \times 5}} = 2{,}117 \text{ fish;}$$
rounded, 2,120 fish.

$$\text{Proportional SE} = 2{,}120 / 11{,}300 = 0.188.$$

With the same data for counts of fish but increasing supplemental information, the resulting proportional standard errors are as follows: for simple random sampling, 0.277;

for mean square successive difference with a series, 0.228; and for ratio estimate based on area, 0.188. Although these values are in the expected order of precision, they represent only this particular data set; they provide no guidance for general use.

To repeat earlier cautions, standard errors calculated for systematic samples are only approximate, possibly useful guides to relative precision and the effect of increasing the sample size.

For any precise determination of confidence limits, or for making direct statistical comparisons between different sets of fish kill data, the investigative plan should be changed to provide unquestionable random sampling, probably multiple (or replicated) systematic sampling with independent random starts, and standard errors calculated without the approximations used here.

Finite Population Correction for Standard Error

When the sample contains an appreciable fraction (more than 10%) of the entire population of unit areas, the standard error may be reduced by a calculated amount. The standard error is estimated and then multiplied by the finite population correction factor:

$$\text{Finite population correction factor} = \sqrt{(1-f)},$$

where f = the proportion of the population that is in the sample.

In the stream example used in Chapter 3 (Narrow Streams, Incompletely Accessible, earlier in this chapter), 4 sample segments were selected from the 12 segments in Stratum II, and the calculated standard error was multiplied by

$$\sqrt{1-\left(\frac{4}{12}\right)} = 0.816.$$

Subsampling

Subsampling means to record data for only part of the fish that are on a sample site. Subsampling can be used to reduce the time and labor of processing dead fish when very large numbers are found in one of the lake or stream segments or transects. With a sample site, either the total area of the site or the total number of fish collected from the site may be subsampled. With a transect, the area over water cannot be easily subsampled, but the collection of fish from it can be.

To subsample the area of a segment, first decide approximately what fraction of the fish will be counted (e.g., one-fifth). Then, divide the sample site into that many subunits (e.g., five subunits). Assign a serial number to each subunit and select the subunit that is to be counted by drawing a random number (Use of Random Number Tables, later in this chapter). Ordinarily, the segment will be divided into fewer than 10 subunits, and a single-digit random number will suffice. Finally, multiply counts from the selected subunit by the total number of subunits in the segment (e.g., five) to expand the number to the whole segment, and use this estimate of the total as though the entire segment had been counted. When a segment is divided into subunits, less sampling error is added if each subunit has about the same number of fish. This condition, however, is not necessary for unbiased sampling, which depends upon random selection of the subunit to be counted.

When it is desirable to subsample a large number of fish that have been already collected from a segment or transect, start with the entire collection. First, decide whether all or only part of the collection is to be subsampled. Perhaps all the larger fish or all the fish except the young of a single species can be processed. Remove and set aside for separate attention those fish that are to be completely recorded. Next, decide what proportion of the remaining fish are to be processed (e.g., one-third) and divide the fish that must be subsampled into the corresponding number (here, three) of small lots (in piles, buckets, or tubs), each containing about the same number of fish. Make all lots as comparable as is practical in numbers, size, and weight of fish by distributing fish among the lots in rotation, not by completing first one lot then another.

If there are too few lots, each too large to count, create more lots by removing some fish from each existing lot. If there are too many lots, distribute the extra fish among the lots to be retained. Assign a serial number to each lot and designate one from a random number table. Weigh the selected lot and all other lots to allow expansion by weight. Count and measure the fish in the selected lot.

To expand the numbers from the selected lot to the total for the segment or transect, multiply the results from the lot either by the number of lots (here, three) or, probably better, by the following factor: total weight of all lots divided by weight of the lot selected for measuring.

Another type of subsampling is the use of differing numbers of sample sites or transects to estimate the numbers of different kinds of fish. For example, in a shoreline kill, trophy game fish may be counted in every third segment (or even enumerated completely) while forage fish may be counted only in every tenth segment. To estimate the total number of all species, add the separate estimates for the different kinds of fish. If a standard error is required for this total, it must be calculated anew from the original counts of all fish made in the least numerous set of sample sites (in this example, data on complete counts on every tenth segment).

Use of Random Number Tables

Random sampling is needed to locate the first counting segment or transect of a systematic sample and to subsample a segment or a large collection of fish from a segment or a transect. A set of the random numbers needed for this purpose is presented in Appendix G.

Selecting a random sample segment in the first gap distance on a stream or lakeshore can be problematic (Segments, earlier in this chapter). For example, in a half mile, there are about nine (8.8) segments of 100 yd each. Number these segments consecutively and select one by drawing from Appendix G a one-digit random number from the range 1–9, with 9 designating the final 100 yd.

Similarly, a sample transect across open water must be located at random within the first gap distance along the survey line. First, determine the number of possible transects in a gap distance—the gap distance divided by transect width. Then, assign a serial number to each possible transect in the gap distance and draw a random number, with the appropriate number of digits, from Appendix G. For example, within a gap distance of 250 yd, there will be about 36 transects of 7 yd width. Select a two-digit number from the range 01–36. Subsamples are selected similarly.

To use Appendix G, first decide the number of digits needed; this will vary with the application. Then, select a starting point on the table by looking away and dropping a finger onto the table. Arbitrarily decide which column or columns to use. Move your finger directly under the closest visible number of suitable size. Then, start with the next number under this, that is, the one concealed by your finger. Progress down the page and read off the next random numbers in sequence as needed, or progress across the page following in the same row.

Significant Digits and Rounding

Significant digits are the reliable digits in a number that are known with certainty, indicating level of precision and accuracy. The same number of significant digits (and implied relative accuracy) appears in 0.12, 1.2, 12, and 12×10^2. In field biology, however, use of exponential 10 is rare and 12×10^2 is usually stated as 1,200. The impact of a good study can be weakened when greater relative accuracy and precision is claimed than exists. Numerical results should be stated in the number of digits justified by the relative accuracy and the number of decimal places justified by the precision of the basic measurements.

The relative precision of a product (or quotient) of a series of quantities is the relative precision of the least precise item. Relative precision is indicated by the number of significant digits to the right of the decimal point that are retained: one digit when measurement of the least precise measurement is within about 10% of the true value, two digits when it is within about 1.0%, and three digits when it is within about 0.1%. Most field measurements have a precision of 0.1–10%. In intermediate calculations, keep one or two digits more than indicated by this rule and round the final figure to the appropriate number of digits.

In estimating numbers of dead fish, the measurements of least accuracy are the linear measurements made to determine area—the length and width of shore segments, for example. Note that numbers of samples and counts of fish are absolute (discrete) numbers determined without a measurement error, even though successive counts are variable. However, the estimated total fish counts calculated from the sample counts are the measurements of least precision. When adding or subtracting numbers, results should be rounded based on the precision of the least precise number. If multiplying or dividing numbers or calculating the square root of a number, results should be rounded based on the accuracy of the least accurate number.

Chapter 3
Field Guidelines for Counting Dead Fish

Chapter 2 provided general background information, and contained instructions for approximating the standard errors of the estimated totals. Chapter 3 provides brief directions for making sampling surveys of fish kills at several sites with the objective of estimating the number of visible dead fish.

Certain advisories in Chapter 2 bear repetition here. English units are used in descriptions of survey procedures, but metric units are equally appropriate. Most procedures are for a one-time, short-term survey, although problems of sampling a kill for more than 1 d are discussed. If a monetary value is to be assigned to the killed fish, Appendix A should be examined to determine whether a length or weight measurement is needed. Because field measurements of fish often are limited to lengths, length–weight conversion functions may be needed. Methods in these chapters were designed for visible dead fish; however, it would be reasonable to utilize these chapters for other dead animals visually observed during a kill. These chapters do not discuss the use of aerial photography, nor do they consider the use of population studies before and after a fish kill to appraise damage.

Sampling principles are the same for segments (streams and shorelines) and transects (open water): dead fish are recorded for each sample site and counts are expanded to estimate total numbers in the study area or in each stratum. For stratified sampling, the totals for each stratum are added to estimate the total number of visible dead fish in the study area. When parts of the area of the kill are inaccessible, the inaccessible area must be set aside as an unmeasurable stratum. Using professional judgment and experience in the accessible areas, the investigator may be able to speculate on the status of the inaccessible areas (Inaccessible Areas, Chapter 2).

Narrow Streams

As used here, a narrow stream is one along which an investigator can traverse each sample site by boat, wading, or walking the banks and can count or collect every visible dead fish along both banks and in the stream channel. A stream too wide for this should be sampled by methods outlined in this chapter for lakes and wide streams. Narrow streams may be either completely accessible or only partly accessible, and guidelines for sampling differ for the two classes.

Narrow Streams, Completely Accessible

When streams are completely accessible for the entire stretch affected by a kill, it is practical to count any segment. To accomplish this, first determine the geographical extent of the kill and record the total length of affected stream. Next, determine the number and location of the sample sites where dead fish will be counted (see Segments, Chapter 2).

If a preliminary trip can be made through the affected section of stream, it may be possible to stake out segments that are varied in length so that each includes about the same number of fish. This practice improves precision (see Precision, Segments, Variability among Sample Sites, in Chapter 2) but at the expense of the extra trip. In this case, record stream length as the total number of segments.

If a preliminary trip is impossible, use digital mapping technology or a map measurer and a map to determine the stream length affected. Next, establish the interval (or gap; Segments, Chapter 2) between the segments to be surveyed (sample sites): a 100-yd sample site every half mile or a 100-m sample site every kilometer throughout the area of the kill is recommended. If a large kill extends over many miles, extend the interval between sample sites. Smaller kills may require smaller sample sites at shorter intervals (e.g., 10 yd every 200 yd). Use at least three sample sites.

When the sample sites have been located, identify and count all fish in each sample site and record lengths, and weights if possible (these values are needed to estimate monetary values of the dead fish). Corrections will be needed if fish are drifting downstream (Narrow Streams with Drifting Fish, Chapter 2). Keep separate field records for each sample site because the data are needed to approximate the standard error (Approximate Standard Error, Chapter 2). Subsample the sample sites that have very large numbers of dead fish (Subsampling, Chapter 2).

Expand the number of fish counted in each sample site to the whole length of affected stream (Estimating Total Number of Dead Fish, Segments, Chapter 2). Either (1) multiply the average count per sample site by the total number of segments or (2) multiply the total number counted in the sample sites by an expansion factor. This factor can be either the total number of segments divided by the number of sample sites or (only if segments are of equal length) the total length of stream affected divided by total length of sample sites. Note: If sample sites vary in length, do not expand the numbers of dead fish with a factor involving length. Use only factors based on number of segments.

Example: Assume that a kill of Bluegills *Lepomis macrochirus* extended over 2 mi (3,520 yd) of a completely accessible stream. The area of the fish kill was divided into four segments. Each segment was 0.5 mi long (gap distance). There are 35.2 potential sample sites (sample segments) in 2 mi (3,520 yd/100 yd = 35.2). The starting location for the first sample site was randomly chosen within the first segment. The 100-yd-long sample site included the entire width of the stream. Three other 100-yd sample sites were selected at successive half-mile intervals. The following numbers of Bluegills were counted:

Length (in)	Sample site 1	2	3	4	Sum
1	140	100	40	0	280
2	120	80	30	10	240
3	60	40	20	5	125
4	30	20	15	10	75
5	25	15	25	15	80
6	30	15	20	5	70
7	10	5	10	0	25
8	5	5	5	0	15
Total	420	280	165	45	910

Estimate the total number of dead Bluegills (all sizes) by either of the following methods.

1. Mean number of fish counted per sample segment calculation:

   Mean number = (910 fish)/(4 sample segments) = 227.5 fish/sample segment.

   In 2 mi, there are 35.2 potential sample segments, each 100 yd long, so

   (227.5 fish/sample segment) × (35.2 potential sample segments) = 8,008 fish; rounded, 8,000 fish.

2. Expansion factor calculation:

   First, calculate the expansion factor in one of two ways:

   Segment count: 35.2 potential sample segments/4 sample segments = 8.80

   Length: [(1,760 yd/mi)(2.0 mi)]/[(4 segments)(100 yd/segment)] = 8.80

   Next, calculate the numbers of Bluegills using the expansion factor:

   910 total number of dead fish counted × 8.80 = 8,008 fish; rounded, 8,000 fish.

   Similarly, if needed, calculate the total number of fish in each length-group. In this example, 240 two-inch Bluegills were collected.

   240 fish × 8.80 = 2,112 fish; rounded, 2,100 fish.

   Round the estimates of dead fish to an appropriate number of significant digits (Significant Digits, Chapter 2).

Narrow Streams, Incompletely Accessible

When faced with a narrow stream that is incompletely accessible, the investigator must adapt the directions given previously for sampling narrow streams. Under these conditions, the accessible parts of the stream should be surveyed with a modified version of the previous directions, as described in this section. For the inaccessible parts of the stream, the investigator may find it appropriate to estimate numbers of dead fish using professional judgment, experience, and observations made in the accessible areas. Such methodology can be useful, and sometimes it is required because it is the best available information. The investigator must, however, make a clear distinction between estimates based on professional judgement and estimates or counts based on surveyed sites.

This approach for surveying incompletely accessible narrow streams is one kind of stratified sampling (Inaccessible Areas and Stratified Random Sampling, Chapter 2).

Stratum I:	portions of the stream under the immediate influence of road-crossing structures.
Stratum II:	accessible portions of the stream beyond the immediate influence of road-crossing structures.
Stratum III:	the inaccessible remainder of the stream (unmeasurable stratum).

Two types of narrow, partly accessible streams are discussed here. For the first type, fish can be counted only in the immediate proximity to road crossings (or other points of easy access, such as fish camps). In the second type, the stream is accessible for varying distances from the road crossings.

Stream accessible only at road crossings (Strata I and III).—Designate a segment of Stratum I immediately at each road crossing and extend the segment upstream and downstream for the distance within which the dead fish can be counted, measured, and weighed. In this case, the limits of Stratum I are set by accessibility, not by an arbitrary distance from the road crossing. At each road crossing, combine into one segment both the upstream and downstream extensions of Stratum I. The upstream and downstream portions at one road crossing should be sampled as one site. This creates a more defensible sampling design. Once an investigator is at the site, it costs little more to count in the second direction. Culverts, bridges, and associated channel or current modifications may cause upstream–downstream differences in the accumulation of dead fish, which would add sampling error if the upstream and downstream areas were counted as separate sites. If possible, count dead fish at all road crossings. If road crossings are too numerous to count all, list the crossings in linear order along the stream and use a systematic sample with a random start to select a subset of crossings (Systematic Sampling with Random Start, Chapter 2).

To estimate the total number of fish for Stratum I, multiply the total number counted at the sample crossings by this expansion factor: total number of crossings divided by the number of sample crossings. *Note: an expansion factor based on length cannot be used if segments are of unequal length.* Keep the counts separate for individual sample crossings if an approximate standard error is to be calculated, (Approximate Standard Error, Chapter 2).

Stream accessible at and beyond road crossings (Strata I, II, and III).—Three or more strata should be used if a stream is accessible at road crossings and for a distance beyond the crossings but inaccessible at greater distances. This stratification retains the advantages of a properly designed sampling survey for Strata I and II. The influences of road-crossing structures are absent in Stratum II; therefore, data from Stratum II provide a reasonable basis for judging the status of Stratum III.

Stratum I is defined here somewhat differently than for the previous case. Stratum I is still located immediately at the road crossings, but here it extends upstream and downstream only far enough to escape possible physical influences of road-crossing structures on the accumulation of dead fish. Subject to the investigator's judgment, a uniform distance of 40–50 yd in each direction is suggested, although this distance can be varied in either direction. At any one road crossing, the extension of Stratum I upstream and downstream should be treated as a combined, single segment, and if selected in the sample, the entire segment should be counted for the reasons stated with the previous case.

In Stratum I, a systematic sample with a random start may be selected from the road crossings listed in stream order; however, all road crossings should be completely enumerated if this is practical. When area sampling must be used, the total number of dead fish in Stratum I may be estimated as the total number of fish counted on all the sample segments in the stratum multiplied by the total number of crossings in the stratum divided by the number of crossings in the stratum sample. Keep the counts separate for individual sample crossings if an approximate standard error is to be calculated for each sample segment (Approximate Standard Error, Chapter 2).

Stratum II includes the accessible part of the stream beyond the immediate influence of road crossings. Stratum II will probably extend for differing distances upstream and downstream at each road crossing, or it may extend in only one direction. The more of the stream that can be included in Stratum II (at the expense of Stratum III), the better the information on total numbers of dead fish will be. If the results for Stratum II are to support speculation about conditions in Stratum III, as suggested here, then the comparative sizes of the two strata must be known.

In Stratum II, it is necessary to set up a sampling frame, which here is a listing in stream order of all segments available for sampling (Area Sampling and Segments, Chapter 2). This list may require a preliminary trip to measure each accessible part of the stream. If so, consider varying the segment lengths so that each contains about the same number of dead fish (Segments, Chapter 2). Stratum II may be enough well known that segments can be delineated on a map without a field exploration.

Select a systematic sample with a random start (Segments, Chapter 2) and count, measure, and weigh the fish in the sample segments. Estimate the total number of fish in Stratum II by multiplying the total number counted on all sample sites by this expansion factor: total number of segments in Stratum II divided by the number of sample segments. Keep the counts separate for individual sample crossings if an approximate standard error is to be calculated (Approximate Standard Error, Chapter 2).

Example: The following example illustrates the use of stratified random sampling with three strata in a narrow stream that was partially inaccessible. The fish kill extended for 4,000 yd (2.27 mi) from a polluting industrial plant to a point downstream where the stream joined a larger tributary and the pollutant was diluted to nonlethal concentrations (Figure 2.4). There were six road crossings along the length of the fish kill. Three strata were delineated, and the total number of dead fish in each was estimated, as follows:

Stratum I consisted of six 100-yd segments of the stream, each at a road crossing and extending for 50 yd upstream and 50 yd downstream from the midpoint of each road crossing. Fish in all segments in Stratum I were counted and measured. The total was 1,906 fish, a value without sampling error. The numbers of fish in individual segments were highly variable; for example, 1,210 fish were concentrated in the 50-yd stretch immediately upstream from one road crossing.

Stratum II consisted of twelve 100-yd segments, one above and one below each road crossing, beyond the Stratum I segments. With 12 segments, ranked in stream order, there are three possible systematic samples, each with four sample segments. One systematic sample of four sample segments was selected at random. The numbers of dead fish on the four sample segments, in stream order, were 30, 47, 115, and 25 fish. The total number of fish in the sample segments (217 fish) was multiplied by an expansion factor of three (total number of Stratum II sample segments, 12, divided by the number of sample segments, 4). The result was 651 dead fish in Stratum II. The standard error (±228) was estimated by mean square successive difference and corrected with a finite population correction factor of

$$\sqrt{1-\left(\frac{4}{12}\right)}$$

(Approximate Standard Error, and Finite Population Correction for Standard Error, Chapter 2).

Stratum III comprised the rest of the stream (2,200 yd), which was remote from roads and impossible to sample. If the investigator judged it reasonable to assume that the density of dead fish was the same in Stratum II and Stratum III, then the number of fish in Stratum III would be roughly in proportion to the lengths of stream in the two strata. Since there were 12–100-yd segments in Stratum II, this suggests about 1,200 fish in Stratum III ([651 fish × 2,200 yd]/1,200 yd = 1,194 fish). The investigator might discount this figure and suggest a lesser number in Stratum III, to be conservative. It is not reasonable to assume that Stratum III had no dead fish.

Using stratified sampling allows a data-based estimate (rounded) of 2,560 ± 228 dead fish in Strata I and II, and a judgment figure of 1,200 more fish in Stratum III, the inaccessible areas.

Accounting for Drifting Fish

Drifting dead fish can create problems. While the investigator is counting and measuring fish in a stream segment, some fish will drift out of the segment, others will drift into it, and still others may float in eddies without moving. If floating fish are few, they can be ignored. Alternatively, floating fish may be moved to the bank and may be added to the shoreline count, resulting in a slight overcount if the investigator is moving upstream while counting or a slight undercount if the investigator is moving downstream. When the current is slow, the floating fish may be added to the shoreline count with little error, regardless of the direction of the investigator's movement.

When drifting fish are abundant enough to require a correction in the estimate of dead fish, the correction will differ with the direction of the investigator's counting path (as explained later). When correcting for drifting fish, keep a separate count of the fish on the bank and those floating (whether visibly drifting or not). Combine the two counts after correcting the number of floating fish for the number of drifting fish. Dead fish drift downstream in an irregular pattern. When counts in the segments are small, the suggested corrections may, by chance, result in zero or negative estimates. An alternative solution is offered for each such case.

To determine the correction when *moving upstream*, count and measure all fish on the stream bank and, separately but simultaneously, count and measure all floating fish, whether drifting downstream or not. Record the time that it takes to finish the segment. Then, for an equivalent time, count all fish drifting past a stationary point at the upstream end of the segment. Correct the number of floating fish counted in the segment by subtracting the number of fish that drifted past the stationary point. If this correction results in a zero or a negative number, use the uncorrected count of floating fish in the sample site. The corrected segment count is the number on the bank plus the corrected number floating.

To determine the correction when *moving downstream*, count the fish on the bank and also keep two counts of floating fish: (1) those that you pass, and (2) those that drift past you. Pay no attention to whether a particular fish has been seen before: count each fish passage. Also record the time required to transit the segment. Then, count all fish drifting past a stationary point at the bottom of the segment for a period of time equal to the transit time. The corrected estimate of floating fish is the number of fish drifting past the fixed point plus the number that you passed minus the number that passed you during the stream transit. If the corrected number is zero or negative, use only the number of fish passing the fixed point. The corrected segment count is the number on the bank plus the corrected number floating.

Lakes and Wide Streams

After a kill in a lake, most of the visible dead fish are found along the shoreline and the rest float in open water. Here, it is more efficient and practical to use two zones (strata), each with a different sampling method (Stratified Random Sampling, Chapter 2), and to sample the shoreline zone with segments and the open-water zone with transects. The total numbers of dead fish are estimated for each zone separately, and then the totals are added.

This method is also appropriate for streams so wide that it is impractical to set up segments and collect fish on both banks and in the open water, as directed for narrow streams. If the shoreline is inaccessible or ill defined (e.g., as with a marsh) or dead fish are not concentrated along the shore, transects may be used for the whole task, without a shoreline zone.

Shoreline Counts

In lakes, it is usual for most of the visible dead fish (often more than 75%) to accumulate along the shore, where they can be surveyed. Set up segments of uniform width in a shoreline zone and select a systematic sample of segments with a random start (Segments, Chapter 2). For a 1-d investigation, extend the shoreline zone into open water far enough to include as many of the dead fish as practical that are visible and easily collected and measured from shore. Once chosen, this width must be used consistently throughout the investigation because it determines where transects must begin and end. If time permits, the shoreline zone can be further divided into two subareas: one for fish stranded on the shore, and the other for fish floating along the shore (Multiple-Day Counts at the Same Location, Chapter 2).

The shoreline zone is surveyed with segments in much the same way as surveys of narrow, completely accessible streams (Segments, Chapter 2). In establishing the segments in the shoreline zone, consider whether it may be practical to vary the length of the segments such that each segment has about the same number of dead fish. Varying the segment length will increase precision for the same number of sample sites but at the cost of the trip required for locating and marking out all the segments before sampling. This cost may reduce the number of segments that can be counted. Estimate the total number of dead fish by methods described for segments. Either multiply the average number of fish per sample site by the total number of segments or multiply the total number of fish counted on all sample sites by the expansion factor: total number of segments divided by the number of sample sites. If an approximate standard error is to be calculated, counts for individual sample sites must be kept separate (Approximate Standard Error, Chapter 2).

Open-Water Counts

The total number of dead fish in the open-water zone (or stratum) is estimated by area sampling (Area Sampling, Chapter 2). The sample site is a transect; transects are parallel strips of known width that run from one shore to the opposite shore, perpendicular to some convenient survey line (Transects, Chapter 2). When used with shoreline sampling, transects must begin and end at the outer edge of the shoreline zone. This requirement prevents double-counting.

Detailed directions for using transects in area sampling are in Chapter 2 (Transects, Chapter 2). Determine an appropriate transect width. Lay out a survey line and select a systematic sample of transects perpendicular to it. Count the dead fish on each transect and keep a separate record for each transect so that a standard error can be approximated, if desired.

If a lake is irregularly shaped, transects can be interrupted by land (Transects 1 and 2, Figure 2.1). In this case, count and measure fish along the portion of the transect located in open water, as though the transect were continuous.

There are two methods of using transects to estimate the total number of dead fish in open water. The appropriate method depends upon the information available. With minimal information and knowing only the length of the survey line, an investigator can use Method A (below). If good maps are available and are adequate for determining both the area of the body of water and the length of each transect, the investigator should use Method B because it increases precision for the same amount of fieldwork.

Example: The hypothetical lake in Figure 2.1 may be used to illustrate both Methods A and B. This lake has an area of about 25 acres and a survey line distance (T) of 1,500 yd parallel to the longest dimension. The transect width (W) is 7 yd, the distance that can be reached from the sampling boat. There are a total of 214 possible transects along the survey line in the study area (1,500 yd/7 yd = 214). Six sample sites are to be run, so the distance between sample sites (the gap distance, measured between transect midlines) is 250 yd. There are about 36 possible transects in the gap distance (250 yd/7 yd = 35.7). A two-digit, random number between 01 and 36 is selected in order to mark the location of the first transect. If that number is 19, it designates the 19th transect of the 36 possible transects in the first gap distance. This sample site is centered about 130 yd from the end of the survey line:

$$(18 \text{ transects} \times 7 \text{ yd}) + (7 \text{ yd}/2) = 130 \text{ yd}.$$

Then, the other five sample sites are located, each centered another 250 yd along the survey line, with the first transect as the starting point. After the transects are perpendicular to the survey line, the starting point of each is marked along the shore to be used as visual reference. There is to be no separate shoreline sampling in this example, so transects would start from the shoreline. The crew in a boat starts from a premarked shore location and proceeds on a compass bearing to the opposite shore, collecting and measuring all dead fish within the width of that sample site. On the six transects, 300 fish in total are collected.

To this point, procedures for Methods A and B have been identical.

Method A, good map not available.—Even without a good map, the direction and length of the survey line must be known as accurately as possible, whether the survey line has been laid out over land (preferably) or over water (if the shore is not sufficiently accessible). Transect lengths, however, need not be measured.

With Method A, the total number of dead fish on the open water of the lake is estimated by using the expansion factor: divide the length of the survey line by the total length of the sample sites. Alternatively, divide the total possible number of transects in the entire study area by the number of sample transects (Estimating Total Number of Dead Fish, Chapter 2).

$$\text{Expansion factor} = T/(W \times N),$$

where

T = survey line length, 1,500 yd
W = transect width, 7 yd
N = number of sample sites, 6

$$\text{Expansion factor} = \frac{1{,}500 \text{ yd}}{7 \text{ yd} \times 6 \text{ transects}} = 35.7.$$

$$\text{Total dead fish} = 35.7 \times 300 \text{ fish} = 10{,}710 \text{ fish};$$
rounded,10,700 fish

If 52 of the 300 counted fish were 6-in Bluegills, the estimated total number of dead 6-in Bluegills would be 35.7 × 52 fish = 1,856 fish (rounded, 1,860 fish).

The standard error of the estimate of 10,700 fish, calculated as an example of the method of mean square successive difference (Standard Error with Mean Square Successive Difference, Chapter 2) is ±2,440 fish (proportional standard error, 0.228).

Method B, good map available.—When lake area and transect lengths can be measured accurately, a more precise estimate of the total usually results from use of the ratio estimate: area of the lake multiplied by the density of dead fish (number of dead fish per unit area). The density is derived as the total number of fish counted on sample transects divided by the product of the transect width and the sum of all sample transect lengths.

For this method to yield improved precision, there are two requirements. First, transects must vary in length, and second, the number of fish per transect must be correlated with transect length (Standard Error with Ratio Estimate, Chapter 2). Ordinarily, these conditions can be expected.

To continue the example, a map of the lake (Figure 2.1) allows the area to be calculated as 1,200,000 yd^2 (247.9 acres) and the aggregate length of the six sample transects as 4,550 yd. Transect widths are 7 yd, so the sample sites have a total area of 31,850 yd^2. The crew collected 300 dead fish on the sample sites, giving a density of 0.009419 dead fish per square yard. Multiplying density by the total area of the lake (1,200,000 yd^2) yields an estimated total of 11,303 fish (rounded, 11,300 fish).

An approximate standard error of the estimated total number may be calculated with a special method appropriate for a ratio estimate. This method uses data on the distribution of the 300 dead fish among the six sample sites and the lengths of the sample sites. Data from this example were used earlier to illustrate the method (Standard Error with Ratio Estimate, Chapter 2). The estimate ± standard error is 11,300 ± 2,120 fish (proportional standard error, 0.188).

Large Meandering Streams

Meandering, wide streams may pose difficult sampling problems that do not seem to match exactly those addressed by previously described methods. Nevertheless, if successive, nonoverlapping segments can be set up (at least on a map) and sampled, it may be possible to adapt the methods for narrow streams that are completely or incompletely accessible (Chapter 2). The boundary lines between adjacent segments should be perpendicular to the current. Designate segments, select a systematic sample with a random start, survey dead fish on the sample sites, and proceed with calculations appropriate for segments (Segments, Chapter 2). The problem of drifting fish may have to be treated differently than it is in narrow streams (Drifting Fish in a Stream, Chapter 2), although the current in some meandering streams may be so slow that this problem is unimportant. Where it is possible to use them, adaptations of narrow-stream methods may be more practical than methods described below because they require less field survey and navigation.

Some meandering streams have well-developed banks, where shoreline methods involving successive sample segments can be used. In other cases, streams are bordered by swamps or marshes, where the edge of open water must be defined arbitrarily, and there is no shoreline accumulation of dead fish.

Transects can be used in two ways to sample the open water of meandering streams. With each method, results will be more precise if a good map is available to allow measurement of the total area of open water and the length of each transect. Each method can be executed without these measurements, but precision will be lower, especially with the first method.

For the first method with transects, lay out a single survey line (as in Figure 2.2), extending over the entire affected course of the stream. This survey line must be just long enough that the perpendicular transects at the two ends of the survey line would just include the spatial limits of the kill. Select an appropriate transect width, determine what the sample size will be, calculate the gap distance, and select a systematic sample with a random start in the first gap distance (see Transects, Chapter 2). The investigator may have to navigate by compass across land to mark the starting point of each transect on the stream and then return later by boat to traverse the sample sites and count and measure dead fish. With this method, some sample sites are likely to be much longer than others (Figure 2.2) and the resulting variability in counts of dead fish will reduce precision (and increase the standard error). If the total area of the kill and the lengths of the sample sites are known, precision can be increased through use of a ratio estimate (Transects, Chapter 2; Standard Error with Ratio Estimate, Chapter 2).

The second method of using transects on a meandering stream is to divide the stream into successive sections, each short enough that one end can be seen from the other (Figure 2.3). Each section must have its own survey line, the length of which must be measured or closely estimated. In the absence of a good map, each survey line must be laid out and measured at the stream. Boundary lines dividing adjacent sections must be defined arbitrarily with landmarks on both banks. All transects must be perpendicular to their respective survey lines, so there is a geometrical problem with any transect placed near the boundary between adjacent sections of a stream. In the triangular area bounded by the boundary line, one shore, and the last complete transect (Figure 2.3), all possible transects are incomplete. Each incomplete transect starts at the shore and terminates on the water at the imaginary boundary line. If an incomplete transect is selected as part of the sample, this endpoint must be located by field approximation. Incomplete transects, if selected, must be used or the resulting estimates will be biased. It may seem tempting to curve the survey line around the stream bend, but the resulting fan-shaped array of transects would sample the inner bend more intensively than the outer bend and result in bias.

When a separate survey line is used for each stream section, two sampling designs are possible. The first design treats each section as a separate survey stratum with one or more transects, and a separate estimate of dead fish is made for each section. These estimates are summed over all sections.

The second design is used when the fish kill is so extensive that some sections cannot be sampled. The section survey lines are listed in sequence on paper as though they formed a single line. The sample sites are then positioned on the new paper survey line as a systematic sample with a random start. Next, the positions for the sample sites are transferred back to corresponding points on the original, single-section survey lines, even if some sections do not have a transect. A single estimate is made for the entire series of sections. Neither this design nor the previously described design escapes the geometrical problem at stream bends.

Multiple-Day Counts

Although this document has concentrated on single-day counts to this point, some investigators may make counts on more than 1 d. Counts restricted to a single day seriously underestimate the total number of dead fish (Counts Underestimate Numbers Killed, Chapter 2; The Nature of Fish Kills, Chapter 2).

At least three problems are associated with multiple-day counts: (1) they are more expensive to investigate than single-day counts; (2) often, the investigator is not present during the first part of the kill; and (3) above all, there are no consistent, agreed upon methods for how counts made over several days should be used to estimate the total number of fish killed. A simple sum of the estimated daily totals for multiple-day counts neglects the problem of counting and recounting some of the same fish day after day.

When multiple-day counts are made, each day's estimate should be based on a separate survey, carried out with approved methods so that each day's result can stand alone. The investigator then may offer a professional opinion, based on experience and knowledge but unsupported by statistical procedure, about the probable grand total for the surveys. Otherwise, there is no obvious simple method of estimating the total number of individual fish killed. Multiple-day counts may be required in two situations: (1) when dead fish continue to appear for several days at the same location, and (2) when a toxic plume of water moves down a stream killing fish at different rates as it proceeds.

Multiple-Day Counts at the Same Location

In most fish kills in lakes, dead fish continue to rise to the surface for several days, where they may be counted. At the same time, scavengers remove or fragment the dead and dying fish, and some fish sink (The Nature of Fish Kills, Chapter 2). Although a single-day count underestimates the number of fish killed, the investigator must judge how much, if at all, the sum of several single-day counts overestimates the total kill.

Professional judgment on the rate of turnover of visible dead fish can be based on the fraction of fish that appear to be fresh (red gills, clear eyes, fresh odor, etc.) and day-to-day changes in the representation of different species and sizes of fish within species.

Under some conditions, a credible estimate might be made of the total number of dead fish stranded on the shoreline over several days. Only counts of fish firmly stranded on the shoreline would be used in this method. Establish segments along the shore with procedures for segments (Segments, Chapter 2), and select and mark the sample sites, which here are to be used day after day. Each day, count, measure, and *remove* all fish that are firmly stranded on the sample sites. Either make expanded single-day estimates and sum them for the multiple-day total or sum the unexpanded daily counts and expand the totals (Estimating Total Number of Dead Fish, Chapter 2). This method assumes, first, that fish do not float off once stranded and, second, that the presence of stranded fish does not affect the rate at which other fish are stranded. Multiple-day estimates of stranded fish cannot be added to the estimates of floating fish because the stranded fish are continually being recruited from the floaters.

Multiple-Day Counts for Kills that Progress Downstream

A plume of water can be so toxic that it kills fish for days as it progresses downstream. Dying fish may move downstream before they become entangled, sink, or strand on the bank. If the investigator waits for the kill to end, some dead fish in the upper stream will be removed by predators and scavengers and others will decompose or drift downstream. The survey of dead fish should start as soon as possible after fish begin to die, and it should follow the kill downstream over successive, nonoverlapping zones (strata) that together cover the area of the kill. Within any stratum, counts should be made at sample sites that are selected at random (Segments, Chapter 2). Stratum estimates may be made as single-day counts or, with suitable procedures, as multiple-day counts at the same location (Chapter 2).

To what degree estimates may be summed across days and across strata depends on the measurement and on the judgment of the investigator. It seems reasonable to sum counts of stranded fish if it can be assumed that they do not move, but not reasonable to sum directly the counts of floating fish across days. Sums across strata may have meaning. In interpreting a multiple-day count, the professional judgment of the investigator at the site is important.

Chapter 4
Costs and Economic Valuation of Fish Kills

Statement of Applicability for Valuation of Fish Kills

The economic guidelines described in this book are based on the concept that the total monetary damage from fish kills is made up of four primary components: replacement costs, biological interim loss value, use value, and nonuse value. The true costs associated with a fish or mussel kill cannot be fully accounted for using replacement cost methods alone. Use and nonuse values should be included as part of monetary damages whenever they can be reliably quantified.

When available, any cost data from hatcheries local to the fish kill site should be used in preference to the general economic data provided in this book. The methodologies described here will remain applicable.

Replacement cost data should not be used for species that are currently listed as threatened, endangered, or otherwise protected at the federal level where specific civil or criminal penalties may apply. To ensure that full compensatory damages and restoration activities are realized, federal and state agencies should consult with each other regarding kills involving federally protected species. In addition to federal claims, many states have threatened and endangered lists and related civil and criminal penalties that may take precedence over the guidelines presented in this document. In some instances, states may be able to claim restitution damages in addition to any federal claims. After some kill events, it may not be possible to restore the site, due to irreparable habitat damage. In these instances, authorities may defer to federal fish kill guidelines.

Government entities frequently seek monetary damages after a die-off of fish due to pollution or other human activities. This chapter assists in calculating those damages. These guidelines are recommendations only and are not intended to supersede any other existing state, provincial, or federal methods for estimating damages after a fish kill. Fisheries vary from one location to another, even within one state or province. Therefore, fishery managers should use their professional judgment and expertise to conduct specific studies, when needed, and to adjust the replacement costs listed in this book, when appropriate, to reflect local fishery conditions and losses.

This chapter provides a standardized economic methodology and sets forth procedures for estimating economic damages to fisheries affected by a fish kill. The methods described in this chapter apply to fisheries that are under the purview of resource management agencies. Commercial fishers and processors normally will apply directly to the legal system for restitution in the case of fish kills.

The information given here is insufficient for evaluating damages from the following types of fish kills:

1. Large-scale kills causing loss of recreational or commercial fisheries or damage to large, heavily used bodies of water;
2. kills in unique aquatic habitats;
3. kills of threatened or endangered species otherwise protected by law; and
4. kills of trophy-sized fish.

When such kills occur, management agencies should conduct special studies to determine how the loss changes peoples' behavior, their use of the resource, and the value they place on the resource. These studies should be based on the concept of "economic value," which quantifies how peoples' values, or well-being, change after major fish kills. If circumstances prohibit such studies, the procedures in this chapter may be used as proxies to calculate damages. However, these procedures usually underestimate legitimate damages in such cases.

The section that follows is an overview of the suggested methodology useful for calculating monetary damages from fish kills. Next, the data required are outlined. Then, the steps necessary to tabulate monetary damages are outlined for five fish kill scenarios. The chapter concludes with a working example of each fish kill scenario. Requisite data are in appendices. Appendix A contains replacement costs of fish. Appendix B gives a list, as of February 24, 2017, of federally threatened and endangered fish to which replacement costs normally would not be applied. Appendix C provides the supporting documentation for the replacement cost data presented in this volume. Appendix D lists examples of legal precedents that support the concept of restitution for fish kills. This list is a sample of cases that have successfully assigned and recovered monetary damages as a result of fish kills. These cases provide conceptual substantiation for the use of the methods presented herein. All monetary expressions are in U.S. dollars. Replacement costs have been standardized to 2016 dollars.

Replacement Costs

Replacement costs include the cost of the replacement fish, the costs associated with restocking the fish kill site and restoring habitats, and the costs of investigating and managing the kill.

Replacement Fish

The replacement costs of fish are based on hatchery production costs. Local hatchery data should be used if they are available; otherwise, the costs in Appendix A may be applied. Replacement costs may be used as proxy costs for killed fish when no other fishery economic information is available, but these minimum costs usually will underestimate substantially the true economic value of the losses.

Replacement data should not be used for species that are currently threatened, endangered, or otherwise protected. The concept of "replacement" does not apply to such species. Further, laws and regulations regarding such species may specify civil or criminal penalties for harm. Fish kills involving protected species typically require extensive interagency coordination (see Chapter 1 for guidance). Replacement data also should not be used for trophy-sized fish except as a last resort. Trophy fish are almost never available from hatcheries, and they have values to fishery users that transcend purchase prices, even if they were available. Defer to user value approaches, described later in this text, when possible.

Restocking and Restoration

Replacement fish must be transported from hatchery to stocking site, and the associated costs of vehicles, fuel, water, prophylactic chemicals, holding containers and nets, personnel, and overhead should be included in replacement costs. If the cause of the fish kill also caused damage to habitats, the costs of chemical and physical rehabilitation—supplies, equipment, personnel, and overhead—should be added as well. Note that state- and federal-protected critical habitats need special consideration, like the protected species that inhabit them.

Investigation and Administration Costs

As documented in Chapters 1–3, investigating a kill, estimating the magnitude of loss, dealing with the public, and coordinating all the activities engendered by a kill are labor-intensive. Relevant personnel costs—which may be incurred by several agencies—should be recovered, as well as the costs of transportation, field equipment, disposal of dead fish, and overhead.

Biological Interim Loss Value

Biological interim loss is a quantification of the biological losses to the fishery for the duration of time starting at the onset of the fish kill to the time the fishery has returned to its pre-kill state or baseline condition (English et al. 2009). This period of time is termed the "interim period." Conceptually, biological loss may be represented by metrics such as lost biomass, lost reproductive output, or other lost ecological services. For simplicity for users of this book, initial costs of fish like those killed listed in Appendix A are used as surrogates for a biological metric. Values per year of recovery are discounted to a present value (see Present Value, later in this chapter) and by a recovery rate. The use of biological interim loss values may be useful for fisheries experiencing heavy to total fish kills that may take many years to return to their pre-kill state. This is especially applicable for fisheries experiencing total kills where restocking is not a viable option and natural recolonization is impeded by fish passage barriers or source populations of fish to recolonize are limited.

When appropriate, investigators may use interim loss for large-scale fish kills but should consider accounting for life history metrics and characteristics of species killed during their assessment.

User Value

User (or use) value is the value of the resource to anglers and others whose use of a fishery will be lost or curtailed by a fish kill. It is generally estimated from the time and money people actually or potentially spend to use a resource such as a fishery. User values are rarely available for a particular fishery before a kill because the necessary studies are laborious. If these values cannot be estimated after a kill, recourse may be made to use data in published economic literature. User values are often expressed as average dollars per person per day. For assessment of damages, user values are multiplied by the number of user-days over the time the fishery is debilitated.

Nonuse Value

Nonuse value is the value people place on a resource's existence or on the opportunity to use the resource in the future. Although the concept of nonuse valuation is documented throughout the economics literature, empirical estimates of fishery nonuse value are few. Nonuse values are thought to be slight for fish kills of small or modest size (the magnitudes for which these guidelines are most appropriate). Due to very limited data, nonuse values will not be considered further in this book.

Caveats

The true costs associated with a fish kill cannot be fully accounted for using replacement cost methods. Hatchery-raised fish may not be of the same quality in terms of survivability, genetic make-up, or ecological services provided compared to wild fish. Use of replacement cost methods may often understate the true resource damages resulting from kill events.

The guidelines in this chapter do not consider long-term behavioral changes of recreational anglers in response to a fish kill, nor are values considered. In addition, losses to charter boat businesses and changes in local economic activity resulting from fish kills are not addressed. Such disruptions or distortions are characteristic of large-scale fish kills. Specialized studies are recommended when economic dislocations are substantial because the costs in this document will not be fully applicable and they will not be sufficient to describe overall damages.

Any economic data specific to the fish kill site or region should be used in preference to the general economic data provided in this book. The general methodologies described here will still apply.

Economic Data

Replacement Fish

The replacement data given in Appendix A are derived from a 2016 survey of public, private, and tribal hatcheries in the United States (see Appendix C for details). This survey provided updates of hatchery production costs published in American Fisheries Society Special Publication 30 (Southwick and Loftus 2003). The current data are stratified (when possible) by geographic region reflecting differences in temperature, water and land availability, and other factors that influence production costs. The regions are large and heterogeneous, so purchase prices or production costs at hatcheries close to the kill site should be used in preference to regional averages if they are known.

Many fish species encountered during a kill are not cultured in hatcheries or reported in Appendix A. For these species, use the replacement costs for the most closely related species given in Appendix A. The hatchery survey did not always yield data for all species in all regions. If appropriate regional data are not found in Appendix A, use the costs provided in the "All" column. These costs are averages of all the data that were obtained.

Within and among regions, not all reporting hatcheries raise the same species to the same size. This accounts for some of the missing data in Appendix A and also for inconsistencies in the rate at which costs increase with fish size within a region.

Production data are lacking altogether for the larger fish of many species. For these, a cost per pound has been calculated based on the relationship between weight and the greatest length for which cost data are available (see Appendix C for details). Because fish grow in worth faster than they grow in size, large fish are undervalued by such weight standards. A conservative approach seems advisable in the absence of information. If weights of killed fish have not been measured in the field, the weight equivalents of large fish can be obtained from length–weight relationships computed by the local fishery agency or obtained from the literature (e.g. Carlander 1969, 1977; Murphy et al. 1991; Schneider 2000).

This publication does not provide replacement costs for species other than fish and freshwater mussels. Other species may also be impacted by a pollution event, and if adequate replacement costs (or other monetary value metrics) are available, then collection of restitution from the responsible party would be justifiable.

Transportation Cost

If the actual cost of transporting replacement fish to the stocking site is not known, an estimate of $1.42 per mile can be used. This is the average cost reported by survey respondents and does not include labor, preparation and loading of transport vehicles, or related costs at the hatchery.

Adjusting for Inflation

Due to inflation, prices charged by fish hatcheries or any manufacturer will change over time. The prices, or replacement costs, provided in this text are based on 2016 fish hatchery costs and prices. It would be reasonable for fish kill investigators to adjust these 2016 prices to reflect the year in which the fish kill occurred using the producer price index (PPI). The PPI "measures the average change over time in the selling prices received by domestic producers for their output" (U.S. Bureau of Labor Statistics 2002). Current PPI information can be obtained from the U.S. Department of Commerce, Bureau of Labor Statistics (www.bls.gov/ppi/). The producer price index for all commodities (PPIACO) has been applied in this publication.

Adjust values annually for inflation using the PPI by calculating the rate of change for a given year

$$RC = (PPI_c - PPI_o)/PPI_o,$$

where RC is rate of change for the current year, PPI_c is the producer price index for the current year and, PPIo is the producer price index for the year the values were calculated.

Then, apply that rate of change to the original value needing adjusting:

$$V_c = (\text{RC} \times V_o) + V_o,$$

where V_c is the fish value adjusted for the current year's producer price index, V_o is the original fish value or the value during the year the values were calculated.

Example: You need to adjust the value of a 1-in Fathead Minnow *Pimephales promelas* for the current year's producer price index.

The value of a 1-in Fathead Minnow in this publication is \$0.04.
The PPI for 2016 (the year the data was collected) is 185.4.
The PPI for the current year is 250.0.

$$\begin{aligned}\text{RC} &= (250.0 - 185.4)/185.4 \\ &= 0.35\end{aligned}$$

$$\begin{aligned}V_c &= (0.35 \times \$0.04) + \$0.04 \\ &= \$0.05\end{aligned}$$

User Value and Contingent Valuation

When at all possible, user values should be recovered after a fish kill. User values (willingness to pay, consumer surplus) can be measured by several means, of which the travel cost and contingent valuation methods are very common (Carson 2012a, 2012b). If user data have not been collected before a fish kill and cannot be obtained soon after, appropriate values may be selected from the economics literature.

User values typically are expressed as average dollars per person per day, per year, or per trip. To obtain the aggregate user value of a fishery, the average-user value must be combined with the number of user-days per year or other relevant time interval lost to the fish kill (such as the time to full recovery of the damaged fishery). Estimates of user days of fishing effort at a fish kill site *before* a kill may be available locally as a result of creel surveys conducted by a fishery agency or university. In their absence, more general estimates may be obtained from the state report series of the most current "National Survey of Fishing, Hunting, and Wildlife Associated Recreation" (USFWS and U.S. Census Bureau). In most cases, estimates of user-days *after* a kill must be estimated by creel surveys or other site-specific observations. The total number of angler-days lost is then multiplied by the value of a representative angler-day. This product should be discounted over time by an interest rate deemed appropriate by the management agency (described in the next section). The result will be the total lost user value from the kill.

In recent years, numerous economic valuation studies have been published. It is recommended that when estimating lost use and nonuse values, investigators should always research the most current valuation studies and also look for values for locations closest to the site in question. A compilation of valuation studies was presented in the sourcebook for AFS Special Publication 24 (AFS 1993), which may be useful for historical or methodological purposes.

Please note that user values alone will not account for all losses resulting from a fish kill. Many species, such as snails, aquatic plants, and other species, are not used by humans but still provide important ecological services and other indirect values, such as forage or habitat for nongame species and improving water quality in the case of aquatic plants. In cases where such physical losses can be quantified, replacement costs can be used as minimal proxy values when available from local hatcheries and other production facilities.

Present Value

The present value is an amount of money that if invested now at a given interest rate, will grow to a target amount by some specified future date. In other words, the present value is today's value of some future dollar amount at a given interest or discount rate.

For example, the present sum (PV) of \$1.00 invested today at a compound interest rate of 4% will grow to the future value (FV) of \$1.125 in 3 years, as determined by the equation

$$FV_t = PV(1 + r)^t,$$

where r = interest rate, or

$$\$1.13 = \$1.00(1.04)^3.$$

Thus, the value of the future sum (FV) of \$1.125 in terms of current dollars or its present value (PV) is \$1.00, as determined by the equation

$$PV = \frac{FV_t}{(1+r)^t},$$

where r = interest or discount rate, or

$$\$1.00 = \$1.13$$

In the general PV formula above, r is typically called the discount rate because to calculate today's value of a future payment or expense, that future payment or expense must be discounted into terms representing today's dollar (PV).

The present value of the stream of lost economic values resulting from a fish kill is calculated using the general formula

$$EV_{pv} = EV_0 + \frac{EV_1}{(1+r)} + \frac{EV_2}{(1+r)^2} + \frac{EV_N}{(1+r)^N},$$

where

EV_{pv} = present value of future stream of lost economic values
EV_0 = economic value in the present year;

EV_1, EV_2, EV_N = economic values in the first, second, and *N*th years after the present year;

r = interest (discount) rate.[1]

For example, suppose lost user value is expected to be \$2,000 per year for 4 years and the discount rate is set at 7% per year. Then,

$$\begin{aligned} EV_{pv} &= \$2{,}000 + \frac{\$2{,}000}{(1+0.07)} + \frac{\$2{,}000}{(1.07)^2} + \frac{\$2{,}000}{(1.07)^3} \\ &= \$2{,}000 + \$1{,}869 + \$1{,}747 + \$1{,}633 = \$7{,}249. \end{aligned}$$

Estimating Monetary Damages

The methods employed to calculate damages vary with the characteristics of a fish kill and with the biological and economic information available. Methodologies are presented in this section to cover five fish kill scenarios. These scenarios are overly simplified for purposes of exposition, and they do not cover all possibilities. In the aggregate, however, they indicate the procedures that can be used—in some appropriate combination—for any fish kill. The procedures do not identify all potential economic losses resulting from fish kills; for example, they do not address nonuse values or disruptions of local economies.

Evaluating the monetary damages from a fish kill should begin as soon as the kill is reported. The extent of the kill and the number of fish lost first need to be measured (Chapters 1–3) because the number of fish killed is required for calculation of monetary loss for most of the fish kill scenarios. Next, the management agency must decide whether to restock or not. Then, the manager can identify the scenario below that best fits the situation and follow the steps given.

Scenario 1: Restocking with Fish Like Those Killed

The fish kill site is immediately restocked with fish identical to those lost in terms of species composition, size distribution, and number. No user-days are lost.

A. Determine the total cost involved with restocking the fish kill site: fish (e.g., see Appendix A), transportation, personnel, and overhead.
B. Add all investigation, rehabilitation, and administration costs, plus any other costs incurred by the managing agency in handling the fish kill assessment and cleanup and rehabilitating the habitat.
C. Add an interest charge if restitution by the party that caused the kill is delayed after agency costs are incurred.

[1] Many states have an official interest rate for such purposes, which should be investigated before deciding which rate to use. In the absence of an official rate, the National Oceanic and Atmospheric Administration recommends a 3% rate in the case of natural resource damages, or defer to such rates associated with 3-month Treasury bills, federal funds rate, or other standard measures of current government interest rates (NOAA 1999).

Scenario 2: Natural Production with User-Days Lost

Natural production is the sole means by which lost fish will be replaced. User-days will be lost.

A. Determine the investigation, administration, and rehabilitation costs. If it is possible to measure and apply user values, proceed with steps B through F. Otherwise, follow the procedures in Scenario 2.1. In this case, the total cost of the fish is the sum of the replacement cost of the killed fish as determined from local hatchery prices or the data in Appendix A.
B. Estimate the time it will take the resource to return to its pre-kill state. Then, estimate the annual number of user-days lost during this recovery period. If the costs of measuring lost user-days are likely to exceed the actual monetary losses, follow the procedures in Scenario 1; if not, continue to C.
C. Multiply the number of user-days lost each year by a user value determined for the affected fishery, for a similar nearby fishery, or by an appropriate value from the economics literature.
D. Calculate the present value (discounted value) of the stream of lost economic values determined in C.
E. Determine the cost of the economic analysis, including contractors and staff time.
F. Add the costs of A, D, and E to arrive at the restitution value.

Scenario 2.1: Natural Production with Biological Interim Loss

This is a variation of Scenario 2 but without any angling pressure; thus, no user-days are lost. Natural production is judged to be the best or only recovery option, but the fishery will not fully recover for a significant period of time after the initial kill.

A. Determine the investigation, administration, and rehabilitation costs. If it is possible to estimate the years it will take the resource to return to its pre-kill state, proceed with step B. Otherwise, follow the procedures in Scenario 1. In this case, replacement cost is the cost of the killed fish as determined from local hatchery prices or the data in Appendix A. Such proxy values probably will underestimate the damages to the resource users.
B. Estimate the years it will take the resource to return to its pre-kill state. If less than 1 year is required or the costs of estimating the years are likely to exceed the actual monetary losses, follow the procedures in Scenario 1. Otherwise, continue to C.
C. Estimate or determine the rate of recovery over the interim period.
D. Calculate the cost of the fish as determined from local hatchery prices or the data in Appendix A.
E. Calculate the present value (discounted value) of the stream of lost values from D and multiply this by the recovery rate (C) to determine the biological interim loss.
F. Determine the cost of the economic analysis, including contractors and staff time.
G. Add the costs of A, D, and F to arrive at the restitution value.

Scenario 3: Restocking with Immature Fish

The fishery is restocked with young fish; the composition of the kill is not mimicked. User-days are lost.

A. Determine the total cost involved with restocking the fish kill site with juvenile fish using the methodology described in Scenario 1.
B. Determine the investigation, administration, and rehabilitation costs. If it is possible to measure and apply user values, proceed with steps C through F. Otherwise, go to step G.
C. Estimate the time it will take the resource to return to its pre-kill state. Then, estimate the annual number of user-days lost during this recovery period.
D. Multiply the number of user-days lost each year by a user value determined for the affected fishery, for a similar nearby fishery, or by an appropriate value from the economics literature.
E. Calculate the present value (discounted value) of the stream of lost economic values determined in D.
F. Determine the cost of the economic analysis.
G. Add the costs of A, B, E, and F to arrive at the restitution value.

Scenario 4: Closure with or without Restocking

The fishery has been closed to all users. Restocking may or may not be done. This is a special case of Scenario 3.

A. Determine the investigation, administration, and rehabilitation costs.
B. If the site is to be restocked, determine the total cost involved with restocking the fish kill site. If it is possible to measure and apply user values, proceed with steps C through F. Otherwise, go to step G.
C. Estimate the time the fishery will be closed, and calculate any additional time (if any) it will take for the resource to return to its pre-kill state. Then, estimate the annual number of user-days lost during the closure and any additional lost during the recovery phase.
D. Multiply the number of user-days lost each year by a user value determined for the affected fishery, for a similar nearby fishery, or by an appropriate value from the economics literature.
E. Calculate the present value of the stream of lost economic values from D.
F. Determine the cost of the economic analysis.
G. Add the costs of steps A, B (if any), E, and F to arrive at the restitution value.

Examples

Fish kill valuation procedures for the five scenarios of the previous section can be illustrated with a simple hypothetical example. A 250-acre lake in Montana supports a sport fishery for Lake Trout *Salvelinus namaycush*. An accidental chemical spill into the lake has dispersed within a few days, causing no permanent habitat degradation, but investigators have determined that 1,000 12-in Lake Trout were killed. The nearest hatchery that can supply Lake Trout is a state facility 50 mi away. It has stocks of both 12-in and 6-in fish;

however, its programs are chiefly experimental, and it does not have relevant production cost records. The following data are required to calculate damages.

Replacement fish. Appendix A provides the estimated cost of replacement fish. Montana is in the Mountain-Prairie region (Figure A.1, page 124) where the average costs for Lake Trout are \$6.21 per 12-in fish and \$2.67 per 6-in fish. Three times as many 6-in as 12-in fish must be stocked to account for mortality as the smaller fish grow to replacement size in the lake. Thus, the replacement costs are \$6,210 for 1,000 12-in lake trout or \$8,010 for 3,000 six-inch fish.

Transportation. The hatchery has not calculated its transportation costs, so the general average of \$1.42 per mile must be used. A 100-mi trip from hatchery to lake and back thus costs \$142 (not including labor). The hatchery will not charge for staff time spent preparing and loading the fish truck, though this is a legitimate reimbursable cost.

Investigation and administration. The cost of the field investigation and associated administration was \$1,000.

Recovery time. The state fishery agency estimates that the fishery will recover immediately if 1,000 twelve-inch lake trout are stocked, in 1 year if 3,000 six-inch fish are stocked, and in 3 years if no fish are stocked.

Angler use. No creel surveys or economic analyses have been done on the affected lake, so fishing effort before the fish kill is unknown and an indirect estimate must be made. Data from creel studies conducted by the state of Montana indicate that approximately 1.8 million days of angling occur annually on the state's lakes and impoundments. The total area of these water bodies is 1.2 million acres, which implies an average fish ing pressure of 1.5 angler-days per acre per year. A 250-acre lake, like the poisoned one, thus should have received about 375 angler-days of use annually before the fish kill.

Loss of angler use. Based on experience elsewhere in Montana, the fishery agency expects that fishing pressure will decline 10%, or 38 angler-days per year, until the fishery has recovered.

User value. A specific user value for the affected fishery is unknown, but in 1996, a value of \$27 per day was estimated for trout fishing across Montana by an independent university study and can be applied in this case. No newer data sources are found in the economic literature. It would be acceptable to increase user values to the current year using the consumer price index issued by the U.S. Bureau of Labor Statistics, not to be confused with the producers price index, which is used to update fish replacement costs. After adjusting for inflation, the \$27 value is now worth \$42 in today's dollar and can be applied to this hypothetical example.

Cost of economic analysis. An economist will spend 1 d, valued at \$500, researching the information needed for an economic analysis.

Interest and discount rate. The state currently is using 5% as a standard interest and discount rate.

Application of these data to the four scenarios follows the same sequence of steps outlined in the previous section.

Scenario 1: Restocking with Fish Like Those Killed (No User-Days Lost)

A. The costs of immediately restocking the lake with 1,000 twelve-inch Lake Trout are \$6,210 for the fish and \$142 for the transportation, or \$6,352.
B. Investigation and administration costs are \$1,000.
C. The company whose truck caused the chemical spill has admitted liability and will pay restitution as soon as the state presents its bill. Thus, no interest will be charged and the total damages (A + B) are \$7,352.

Scenario 2: Natural Production

A. Fish will not be stocked and there will be no habitat rehabilitation. The state's initial cost is \$1,000 for investigation and administration.
B. A loss of 38 angler-days per year is expected over the 3 years needed for the fishery to recover.
C. Multiplication of 38 lost angler-days by the user value of \$42 per day yields an annual lost value of \$1,596.
D. Discounted 5% for 3 years, the annual lost angler value of \$1,596 has a present value of \$1,596 + (\$1,596/1.05) + ($\$1,596/1.05^2$) = \$4,564.
E. The cost of the economic analysis is \$500.
F. The total damages (steps 1 + 4 + 5) are \$6,064.

Scenario 2.1: Natural Production, No Angler Days Lost: Biological Interim Loss

A. Fish will not be stocked and there will be no habitat rehabilitation. The state's initial cost is \$1,000 for investigation and administration.
B. Without restocking, it is expected full recovery of the fishery will take 3 years, assuming a linear recovery rate.
C. If stocking had been a viable option, the costs to restock fish like those killed is \$6,210 (annual biological interim loss value).
D. Discounted 5% for 3 years with 33% recovery during year 1 and 67% recovery during year 2, the annual biological interim loss value of \$6,210 has a present value of \$6,210 + ([\$6,210/1.05]*0.67) + ([$\$6,210/1.05^2$]*0.33) = \$12,032.
E. The cost of the economic analysis is \$500.
F. The total damages (A + D + E) are \$13,532.

Scenario 3: Restocking with Immature Fish

A. The total cost of immediately restocking the lake with 3,000 six-inch Lake Trout is \$8,010 for the fish and \$142 for transportation, or \$8,152.
B. Investigation and administration costs are \$1,000.
C. A loss of 38 angler-days is expected over the 1 year needed for the fishery to recover.
D. Multiplication of 38 lost angler-days by the user value of \$42 per day yields a total loss value of \$1,596.
E. No discount is applied during the year after the kill, so lost angler value is \$1,596.
F. The cost of the economic analysis is \$500.
G. The total damages (A + B + E + F) are \$11,248.

Scenario 4: Closure with or without Restocking

A. Investigation and administration costs are $1,000.
B. The total cost of restocking the lake with 1,000 twelve-inch Lake Trout is $6,210 for the fish and $142 for transportation, or $6,352.
C. The fishery will be closed for 1 year. If replacement fish are stocked, the fishery will require no additional recovery time, but a loss of 375 angler-days will result from the year of closure. If the fish will not be restocked, two more years will be needed for the fishery to fully recover, and 38 additional angler days will be lost for each year of closure.
D. Multiplication of 375 lost angler-days by the user value of $42 per day yields a total loss value of $15,750 for the year of closure. During the two recovery years, if fish are not stocked, the loss value will be 38 × $42, or $1,596 per year.
E. No discount is applied during the year of closure, so the lost angler value is $15,750 if fish are restocked. If fish are not restocked, lost value is discounted at 5% for the two recovery years after closure. The present value of loss in this case is $15,750 + ($1,596/1.05) + ($1,596/1.05^2), or $18,718.
F. The cost of the economic analysis is $500.
G. If the lake is restocked, total damages are $23,602 (A + B + E + F); if the lake is not restocked, total monetary damages are $20,218 (A + E + F).

It is readily apparent from these examples that the way in which a fish kill is valuated (with replacement or lost-use values) can result in widely differing costs. It is also apparent that good information is important but costly.

The most efficient approach, defined as the approach accurately quantifying the greatest net damages (total monetary damages minus the cost of investigating and evaluating monetary damages), should be selected when alternative strategies are possible.

Chapter 5

Procedures for Investigation of Freshwater Mollusk Kills[1]

Goal and Coverage of this Chapter

The goal of this chapter is to provide guidance to resource professionals to assist with the entire investigative process, from first knowledge of a mollusk kill incident to the kill assessment to preparation of a final report.

This chapter is divided into several sections. The first section provides background information on freshwater mollusks and an overview of kill investigations. In the second section, useful forms for the kill investigation are provided. The final section details procedures for investigating a freshwater mollusk kill, including initial notifications, what measurements to take, quantifying the magnitude of the kill, and preparing the final report. Chapter 6 provides methods for making a quantitative estimate of the total mollusk kill. This chapter draws from the fish kill guidance manuals developed by the American Fisheries Society (Southwick and Loftus 2003) and Meyer and Barclay (1990) and makes use of many of the same procedures, recommendations, and forms provided in those publications.

Importance of Mollusks

Freshwater mollusks, including mussels (Unionoida) and snails (Gastropoda), are widely distributed across North America. North America supports nearly 300 recognized species or subspecies of mussels (Turgeon et al. 1998; Williams et al. 2017) and more than 700 snail species (Johnson et al. 2013). Freshwater mollusks occupy a prominent niche in a variety of freshwater ecosystems, including fast-moving mountain springs, sluggish coastal streams, small creeks, large rivers, ponds, lakes, reservoirs, and wetlands. They are recognized as excellent indicators of water and habitat quality due to their sensitivity to many types of environmental perturbations. Mollusks provide a food source for several fish species, as well as aquatic and riparian mammals, birds, and reptiles. Snails have a profound impact on algal primary productivity, playing a pivotal role in nutrient cycling, and mussels remove large amounts of suspended solids and contaminants from the water column through their filter-feeding activity. Even the shells of mussels serve a purpose in the ecosystem, providing substrate and cover for invertebrates and cover and spawning sites for some small fish species. The commercial history of mussels in the button industry has been well documented (Thiel and Fritz 1993; Anthony and Downing 2001), and mussels continue to possess significant economic value in the international cultured pearl industry and in bead production.

[1] Prepared under the guidance of the Freshwater Mollusk Conservation Society.

Freshwater mollusks have been declining in diversity and abundance in North America over the past 200 years, to the point where the percentage of extinct, endangered, or threatened species is far greater than that of any other animal group (Johnson et al. 2013; Haag and Williams 2014). Most of this decline has been associated with the ever-increasing encroachment of humans upon aquatic habitats in the form of impoundments, sedimentation, channelization and dredging of rivers, illegal harvest, and aquatic contaminants. The invasive Asian clam *Corbicula fluminea*, zebra mussel *Dreissena polymorpha*, and quagga mussel *D. bugensis* have also had a significant impact on native mollusks. The unique reproductive cycle of freshwater mussels, in which a specific fish host is required for a mussel's parasitic larval stage, has undoubtedly played a significant role in the decline of many species, as the same environmental perturbations impacting mussels also impact the availability of the host fish species.

Causes of Kills

As with the better-publicized fish kills, freshwater mollusk kills can be caused by a wide variety of agents, both natural and manmade. Past examples of identified or suspected causes include the release or spilling of toxic chemicals, untreated domestic sewage or waste from livestock operations, agricultural runoff following pesticide applications, depletion of dissolved oxygen (natural or human-induced), elevated water temperature, bacterial or viral infections, parasitic infestations, and invasive species.

Too often, however, mollusk die-offs have not been linked to a specific cause. A rapid, coordinated response (e.g., knowing who to notify and knowing what measurements to take) should improve our ability to make such links between cause and effect. The U.S. Environmental Protection Agency (USEPA 2000a) published its "Stressor Identification Guidance Document," which is a useful tool to identify the causes of observed kill incidents and to provide a structure for organizing the scientific evidence supporting the conclusions of the investigation. The guidance outlines a five-step process for conducting a causal assessment at a particular site: (1) define the case, (2) list candidate causes, (3) evaluate data from the case, (4) evaluate data from elsewhere, and (5) identify probable causes.

Statistics on Kills

Table 5.1 provides recent examples of the numbers of mussels involved in freshwater mollusk kills. Although far from complete, this information serves to highlight the seriousness of the problem, particularly when one considers the documented global imperilment of the resource. Also, note that these figures are exclusively for mussels. Snails have typically not been evaluated in past kill assessments. Given the similarly widespread endemism, rarity, and intolerance to environmental perturbations, snails should certainly be accounted for, whenever possible, in future assessments.

Difficulties Commonly Encountered

The first difficulty encountered in the investigation of mollusk kill incidents is detecting the kill in the first place. Dead mussels and snails are typically not as visible as dead fish.

TABLE 5.1. Examples of selected mollusk kill statistics.

Location	Date	Cause	Kill estimate	Fish kill also?
Clinch River, VA	1998	Chemical spill	18,000	Yes
Ohio River, OH and WV	1999	Chemical release	1,000,000+	Yes, primarily Freshwater Drum *Aplodinotus grunniens*
Big Darby Creek, OH	2000	Chemical release		Yes
Dunkard Creek, PA and WV	2009	Chemical release, toxic algae bloom	45,000+	Yes (and mudpuppy *Necturus* sp.)
Yellowleaf Creek, AL	2006	Chemical release, toxic algae bloom	250+ mussels, 1 million+ snails	Yes
Roebuck Spring Run, AL	2008	Removel of low level dam resulting in habitat destruction	1 million+ snails	Yes
Roebuck Spring Run, AL	2013	Removal of low level dam resulting in habitat destruction	350,000+ snails	Yes

Often, dead mussels will be found buried in the substrate in the same position they occupied when alive, whereas fish typically become visible floating at the water surface or washed up on the shore. While it should become routine to investigate the possibility of an associated mollusk kill whenever a fish kill is reported, it is not unusual for mollusks to be impacted when the impact to fish is negligible or nonexistent. Mortality of freshwater mollusks, particularly mussels, may not occur immediately, as most species have the ability to "clam up" for extended periods of time. However, use of energy reserves during this period could affect future survival, and mortality may not be observed for several months. The timing of the kill may also obscure evidence; a spill occurring in colder months may not produce a mussel kill until the following spring when mussel metabolism increases due to warmer temperatures. Periodic follow-up may be necessary to detect a kill event.

Accessibility and safety are also key aspects of kill investigations. As with any field sampling, access to remote or privately owned areas can present obstacles to effective sampling of a kill incident. This problem is most often associated with small streams. Use of a boat or canoe may be necessary to reach some areas. In addition, if a large river or lake needs to be assessed for a possible kill incident, certified divers will probably be required, adding time and expense to the investigation. Safety issues associated with the source of the kill may also limit the investigator's ability to conduct a search for dead mollusks. An assessment may be hampered due to immediate unknown causes of the kill and thus limit investigator's contact with water until the source of the kill is known and the waters are determined to be safe for human contact. A related issue is being able to effectively sample the impacted assemblage of mollusks. Individuals may be in deepwater habitats, buried in substrate, or be in other difficult-to-sample areas. If baseline information on the location of mussel assemblages is not available, locating impacted or reference assemblages may be difficult in some areas. Snails may be difficult to locate because of their small size and corresponding susceptibility to being carried off by fast-moving currents. Additionally, when making counts, the investigator must be able to distinguish between fresh dead[2] and relic mollusk shells.

A basic difficulty encountered is knowing what information to collect. When faced with a recent or ongoing kill incident, the investigator must be able to quickly discern which environmental and biological information is relevant. Often, available evidence may initially point to several possible causes. The investigation requires careful observation, accurate record keeping, and proper selection of laboratory analyses. Additionally, typical limitations on the time and resources available to devote to an investigation dictate efficient decision making in selecting what information is needed.

An important aspect of these investigations is maintaining standards of legal evidence. Engaging in successful legal action against the entity responsible for a kill requires a timely response and strict attention to quality assurance and quality control (QA/QC), such as proper recording of observations, collection of appropriate samples, performance of necessary tests, maintenance of chain-of-custody procedures, and proper documentation of evidence.

A factor that cannot be ignored in these investigations is the ongoing lack of public awareness of the peril, sensitivity, and importance of freshwater mollusks. To increase the likelihood of appropriate outcomes from legal actions related to kills, and to reduce the frequency of kill incidents, public education on the value and status of these organisms is needed.

Planning the Investigation

Forms and Checklists

The following pages provide examples of forms and checklists that may be useful in an investigation of a mollusk kill; it is recommended that state and federal agency malacologists be consulted for specific forms to use in each state. An equipment checklist (Form 5.1) is a must for any field sampling, and kill investigations are no exception. Also included are a notification form (Form 5.2), field observations/habitat assessment form (Form 5.3), specimen information form (Form 5.4), evidence tag (Form 5.5), kill enumeration form (Form 5.6), chain-of-custody form (Form 5.7), and report form (Form 5.8). These forms represent typical types of information that should be collected; when faced with a kill,

[2] "Fresh dead" throughout the publication refers to shells with clean and shiny nacre on the inside of shells. Weathered dead refers to shells with chalky nacre on the inside.

FORM 5.1. **Example of suggested equipment checklist.**

Some or all of the following equipment will be needed based on local conditions.

- Phone numbers of cooperating agencies
- Emergency contact
- Hospital information
- List of available sources for analytical or diagnostic services
- First aid kit
- Global Positioning System unit
- Compass
- Area map
- Measuring tape
- Flagging
- Rope
- Stopwatch
- Measuring board or calipers
- Cooler with ice
- Plastic bags with labels and fasteners
- Water sample containers (glass, plastic)
- Glass vials with screw caps
- Preservatives (i.e., formalin, alcohol, HNO_3, H_2SO_4, NaOH, zinc acetate, sodium thiosulfate), as needed
- Bacteriological sample tubes and related items
- Shipping boxes, packing materials/labels
- Aluminum foil
- Paper towels and garbage bags
- Field notebook
- Pencils and waterproof marking pens
- Clipboards
- Waterproof forms:
  - Initial notification form
  - Field observation/habitat assessment forms
  - Specimen information forms
  - Chain-of-custody forms
  - Kill enumeration forms
  - Investigation report forms
- Tape recorder and video recorder
- Flashlights
- Taxonomic keys
- Mollusks kill field manual
- Rain gear
- Gloves, boots
- Shovel, tubs, buckets
- Respirators with appropriate cartridges
- Dip nets
- Stainless steel scoop
- Quadrats
- Dissecting pans and tools
- Sieves
- Soft-bristle brush
- Penknife
- Forceps
- Compound microscope and accessories
- Reverse pliers
- Tally meters
- Calculator
- Collecting bags
- Underwater viewing scope
- Water quality meter capable of measuring dissolved oxygen, temperature, pH, and conductivity
- Secchi disk
- Kemmerer sampler
- Dredge sampler
- Core sampler
- Rake
- Boat, life preservers, and related items
- Diving equipment
- Mask and snorkel
- Two-way radios

investigators should defer to their agency-specific forms and procedures if they exist. The content of the form that an investigator actually uses can vary based on the state's protocol and specific situation. Some of the redundancy across the forms may be reduced in the final set of forms that a particular agency adopts and all forms may not be needed depending on circumstances.

Form 5.2. Example initial mollusk kill notification.

STATE: ______________ DATE OF KILL: __________ DATE REPORTED: ______________

TIME REPORTED: ________________ TIME SINCE INCIDENT (if known): __________

NAME OF REPORTER: ____

ADDRESS: ____________________________________ PHONE: ______________

ORGANIZATION: ____

WATER(S) INVOLVED: ____

STATE: _________________________ COUNTY: _______________________

SPECIFIC LOCATION OF KILL (bridge, highway, state road, landmark, etc.):

____

RIVER MILE: _____________ COORDINATE SYSTEM: ___________________

Coordinates Upstream: ____________________ Downstream: _______________

SUSPECTED REASON FOR KILL (natural versus human-induced and possible source of discharge/spill/runoff): ____________________________________

____

____

INITIAL REPORT OF SPECIES INVOLVED: ______________________________

____

____

ESTIMATED NUMBER DEAD: ________ MOLLUSKS STILL DYING?_____________

PERSONS AND AGENCIES NOTIFIED:

NAME	TIME	PHONE

ADDITIONAL COMMENTS:

REPORT RECEIVED BY: ____

DATE: ____

Form 5.3. Example field observation/habitat assessment form.

Page 1 of 2

DATE OF KILL: ________________ TODAY'S DATE: ________________

INVESTIGATOR NAME/ORG: ________________________________

EXTENT OF KILL:	LOCATION OF THIS ASSESSMENT:
Waters(s): ____________________	Water(s): ____________________
County: ____________________	County: ____________________
State: ____________________	State: ____________________
Description (attach map):	Description (attach map):

MOLLUSK CONDITION:

1. ( ) Partial ( ) Complete kill ( ) Undetermined kill ( ) Ongoing kill
2. Status of kill: ( ) Animals dead, soft parts present
   ( ) Animals fresh dead, soft parts absent
   ( ) Many animals still dying
3. Species affected: ____
4. Life stages affected: ( ) <1 year ( ) ≤5 years ( ) >5 years
5. No. affected: ( ) <100 ( ) 100–1,000 ( ) 1,000–10,000 ( ) >10,000

CONDITION OF OTHER BIOTA (check all that apply):

Organism group	Kill (check)	Species involved	Comments
Algae			
Macrophytes			
Zooplankton			
Insects			
Fish			
Crayfish			
Other vertebrates			
Invasive mollusks			

FORM 5.3. Continued.

ENVIRONMENTAL CHARACTERISTICS

NOTE: Investigators should specify units of measure for all parameters.

WEATHER: Air temperature: ________________ % cloud cover: ____________

Wind speed/direction: ____________ Precipitation: ____________

WATER MEASUREMENTS:

Location	Time	pH	Clarity	Color	DO	Cond.	Temp.	Current	Other

River flow condition: ( ) low ( ) moderate ( ) flood

Identify and list nearest USGS GAGE readings ______________________________

HABITAT/RIPARIAN CONDITION:

Bottom sediments:

Compaction: ( ) firm ( ) medium ( ) soft

Predominate substrate type(s): ______________________________

Sedimentation/deposits present: ______________________________

Land use (kill site):

( ) forest
( ) old field
( ) row crop
( ) pasture
( ) wetland
( ) industrial
( ) urban
( ) residential/park
( ) mining/construction
( ) other: ____________

POTENTIAL IMPACT SOURCES (describe):

( ) discharge pipe ______________________________

( ) runoff ditch ______________________________

( ) habitat disruption ______________________________

( ) spill occurrence ______________________________

( ) other ______________________________

NOTES:

____

FORM 5.4. Example specimen information form.

SPECIMEN ID NO: ______________________ DATE: __________

PROCESSED BY ______________________________

Organization: ______________________________

CHAIN OF CUSTODY: ( ) Yes ( ) No

SINGLE SPECIMEN: ( ) OR MULTIPLE SPECIMENS ( )

SPECIES: ______________________________

Length: ______________ Life stage: ( ) <1 year ( ) ≤5 years ( ) >5 years

Source: ______________________________

GROSS EXTERNAL EXAMINATION:

SHELL: ( ) normal ( ) fresh abrasion ( ) crack ( ) puncture

( ) encrustation by ______________________________

GROSS INTERNAL EXAMINATION:

SHELL: ( ) fresh; clean, shiny nacre ( ) weathered

SOFT PARTS: ( ) absent ( ) normal ( ) abnormal color: ______________

( ) parasites ( ) reduced ( ) distended

REPRODUCTIVE CONDITION: ( ) juvenile ( ) female gravid

( ) female not gravid ( ) male ( ) unknown

Other/comments: ______________________________

DISPOSITION OF SPECIMEN:

( ) Shipped for testing: ( ) shell ( ) soft parts

( ) Saved for evidence: ( ) shell ( ) soft parts

( ) Other: ______________________________

TESTS REQUESTED:

( ) Tissue residue: ( ) metals: __________ ( ) organics: __________ ( ) other: __________

( ) Biomarkers: ( ) metallothionein ( ) cholinesterase ( ) other: ______________

( ) Bacteriology ( ) Virology ( ) Histology

( ) Toxicity: ( ) water column ( ) sediment

FORM 5.5. Evidence tag.

FRONT

DATE COLLECTED ________ TIME ______

SAMPLE ID NO. ________________________

TYPE OF SAMPLE: Mussel _____ Snail _____ Other _____

SAMPLE LOCATION ____________________________________

SAMPLE PRESERVED WITH ________________ _____NONE

OTHER INFORMATION _________________________________

COLLECTED BY: ______________________________________

BACK

SAMPLE TRANSFER RECORD – SIGNATURES REQUIRED

DATE/TIME

FROM ________________ TO ____________________________

FROM ________________ TO ____________________________

FROM ________________ TO ____________________________

FROM ________________ TO ____________________________

FROM ________________ TO ____________________________

FORM 5.6. Kill enumeration form.

NOTE: Investigators should specify units of measure for all parameters.

TYPE OF SAMPLE: ( ) Transect (length: ________; width: ________) DATE: ________
( ) Timed search (description of area: ________)
( ) Quadrat (size: ________)
( ) Other (describe: ________)

LOCATION: Water: ________ River mile/location ID: ________
TIME: Start: ________ Finish: ________ Coordinate system: ________
Coordinates upstream: ________ Downstream: ________
INVESTIGATORS: ________

	Count						
	Adults (>5 years)			Juveniles (≤5 yrs old)			
Species	**Live**	**Fresh dead**	**Weathered shell**	**Live**	**Fresh dead**	**Weathered shell**	**Disposition***

* Returned to source, bagged as evidence, kept for further analysis, etc.

Form 5.6. Continued.

COLLECTOR'S NAME AND AFFILIATION:	DESCRIPTION OF SHIPMENT:	
	NUMBER OF SAMPLES:	
	NUMBER OF CONTAINERS:	

SAMPLE ID NUMBER	SAMPLE DESCRIPTION	COLLECTED		ANALYSES REQUESTED FROM LABORATORY	PRESERVATIVE
		DATE	TIME		

Relinquished by (signature)	Date	Time	Received by (signature)	Carrier (If shipped)	Send report to
Relinquished by (signature)	Date	Time	Received by (signature)	Carrier	
Relinquished by (signature)	Date	Time	Received by (signature)	Carrier	

FORM 5.8. Example mollusk kill investigation report form.

Page 1 of 5

Case No. __________ Date: __________
Prepared and submitted by ____________________

1. Kill reported to investigator by
   Name ____________________
   Address ____________________
   Phone __________ Date __________ Time __________
2. Investigator reported kill to
   Name ____________________
   Address ____________________
   Phone __________ Date __________ Time __________

   Name ____________________
   Address ____________________
   Phone __________ Date __________ Time __________
3. Body of water ____________________
4. Nearest town __________ County __________ State __________
5. Specific location of kill ____________________
6. Time and date of kill ____________________
7. Time and date of investigation ____________________
8. Suspected cause of kill, or describe how kill occurred: ____________________
   ____
   ____
9. Duration of kill ____________________
10. Extent of kill: Stream miles __________ Lake acres __________
11. Estimated total number of mollusks killed (from #24) __________ Value $__________
12. Condition of mollusks found (check one or more):
   ( ) Alive but gaping (shells slightly open) ( ) Fresh shells, tissue present
   ( ) Fresh shell, w/o tissue present ( ) Weathered shells

Form 5.8. Continued.

Page 2 of 5

13. Other organisms affected: ____________________

14. Associated water/sediment/habitat conditions of note ____________________

____

15. Investigation personnel (give name and organization represented):

____

____

____

____

____

16. Other persons and organizations involved:

Name and title	Organization and address
____________________	____________________

Remarks ____________________

____________________	____________________

Remarks ____________________

____________________	____________________

Remarks ____________________

17. Topographic view or map of the affected area:

Form 5.8. Continued.

NOTE: Investigators should specify units of measure for all parameters.

18. Chemical/physical measurements: Field

Sample location	Sample type*	Sample depth	Date	Time	pH	DO	Cond.	Temp.	Other	Other	Other

* Water, sediment, etc.
Collected by ______________________________

19. Chemical/physical measurements: Laboratory

Sample location	Sample type*	Sample depth	Date	Time	Parameter	Parameter	Parameter	Parameter

* Water, sediment, etc.
Analyzed by ______________________________

20. Discussion:

21. Recommendations for corrective action needed:

Form 5.8. Continued.

Note: Copy this form if needed for additional species.

22. Projected loss of mollusks:
    Water body __________
    Stream length: __________ Stream width: __________
    Lake surface acres __________
    Total length of stream/lake affected (include units)__________
    Sampling method __________
    Total kill estimation method (count, extrapolation, etc.)__________

SPECIES__________

Age	Count	Est. total	Value/ mollusk	Total value		Age	Count	Est. total	Value/ mollusk	Total value
0						10				
1						11				
2						12				
3						13				
4						14				
5						15				
6						>15				
7										
8										
9						Subtotal				

SPECIES__________

Age	Count	Est. total	Value/ mollusk	Total value		Age	Count	Est. total	Value/ mollusk	Total value
0						10				
1						11				
2						12				
3						13				
4						14				
5						15				
6						>15				
7										
8										
9						Subtotal				

FORM 5.8. Continued.

Page 5 of 5

SPECIES____________________

Age	Count	Est. total	Value/ mollusk	Total value		Age	Count	Est. total	Value/ mollusk	Total value
0						10				
1						11				
2						12				
3						13				
4						14				
5						15				
6						>15				
7										
8										
9						Subtotal				

Total: ____________________

Prepared by ____________________

Title: ____________________ Date: __________

23. Field investigation expenses:

Agency ____________________

Name	Hours	Salary	Miles driven	Mileage rate	Boat operation costs	Other (itemize)

Prepared by ____________________

Title: ____________________ Date: __________

24. Totals:

Estimated total number of mollusks lost: __________

Estimated total value of mollusks lost: $__________

Agency expenses: $__________

GRAND TOTAL $__________

Maintenance of Equipment

Advanced preparation is key to a successful field operation, and part of this is making sure that equipment is properly stored and maintained so that it will be ready when needed. A sample equipment checklist is provided (Form 5.1), and it is recommended that a regular maintenance check procedure be adopted specifying periodic checks of essential items, such as those that are battery-operated, with maintenance logs. Maintenance checks should also include equipment with renewable supply items, reagents for analytical test kits, and other types of chemical and biological samples. Another useful item is to maintain topographic maps or other geographic maps of the region as backups to a Global Positioning System unit or software on a mobile device.

In addition, certified divers and diving equipment may be required to investigate deepwater habitats. If the investigating agency does not possess this capability in-house, arrangements should be made to have this service readily available when needed. In some cases, remote underwater video may be an acceptable substitute for initial reconnaissance of deepwater habitats depending on water clarity.

Procedures for Investigating a Mollusk Kill

Who to Notify

When responding to a possible kill incident, as much information as possible should be gathered from the reporting party, including the suspected cause of the incident. A mollusk kill notification form (Form 5.2) is a useful tool in the initial response to an incident. As soon as possible in the process, a case number should be assigned to the investigation to be used on all tags, data sheets, and other records.

A list of agency contacts (and telephone numbers) to be notified in the event of a kill should be maintained. Such a list will typically include

State water protection agency
State fish and wildlife agency
U.S. Fish and Wildlife Service
U.S. Environmental Protection Agency

In cases involving spills of oil or hazardous materials, additional agencies will need to be notified:

National Response Center, Washington, D.C. (U.S. Coast Guard)
State emergency response agency
Local emergency response agencies

Response Coordination

Because of the need for quick action when responding to a possible kill incident, and the number of federal, state, and local governmental agencies that may need to be involved or notified and kept informed of the investigation, it is essential to plan the coordination of the investigation in advance to the extent practical. Figure 5.1 provides a flow chart depicting the main steps in the process. Notification of appropriate agency personnel early in the

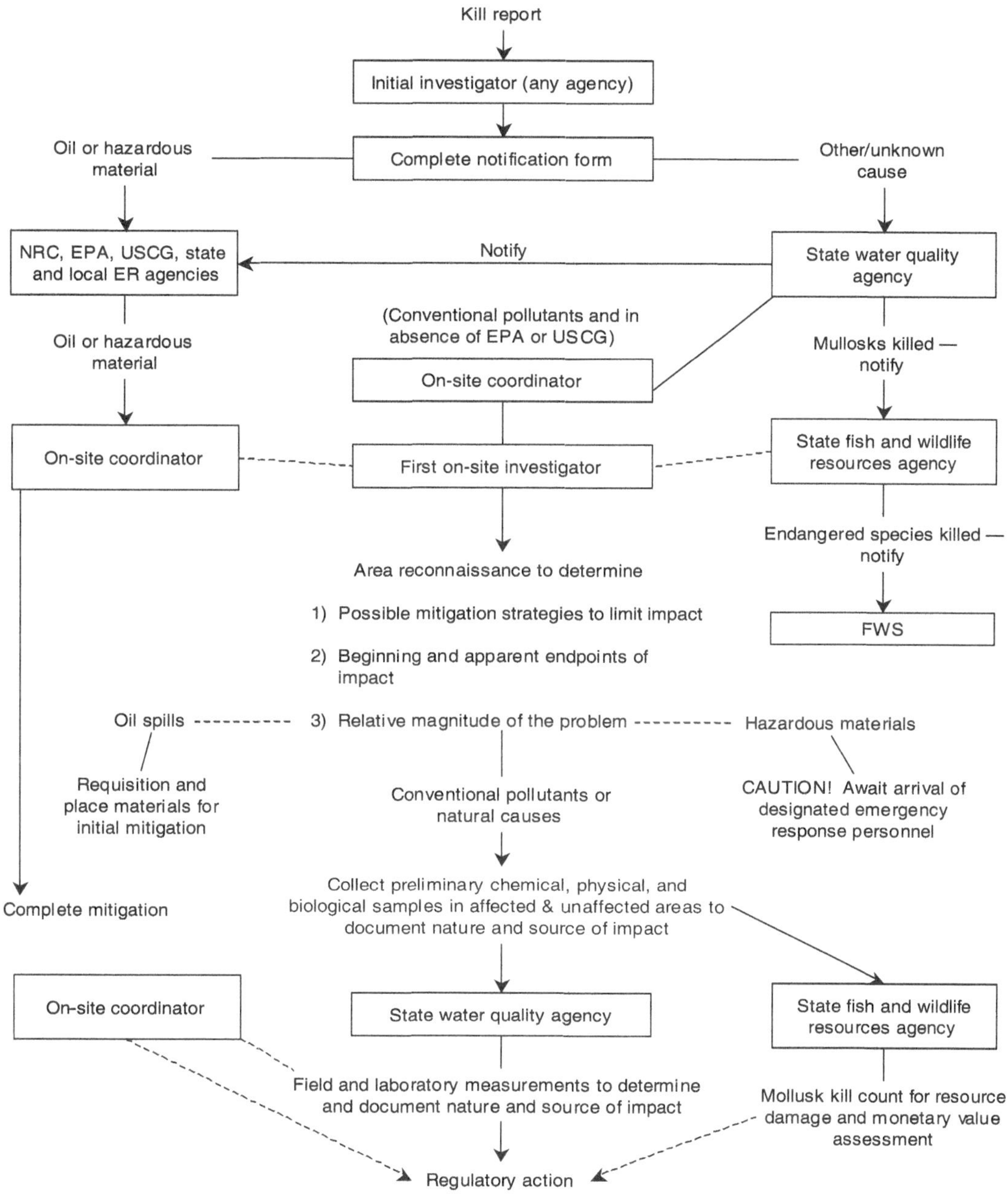

FIGURE 5.1. Flow chart for coordinating a mollusk kill investigation (modified from Meyer and Barclay (1990). Abbreviations: EPA = U.S. Environmental Protection Agency, ER = Emergency Response, FWS = U.S. Fish and Wildlife Service, NRC = National Response Center, and USCG = U.S. Coast Guard.

process will serve to initiate the communications network among participating parties. State and local personnel will often have familiarity with the kill locale, which can prove valuable for determining possible causes, sample locations, and other key decisions. It is important to keep all participants fully informed by providing regular updates on the progress of the investigation. Doing this will improve both the effectiveness and ultimate acceptability of the investigation to all parties.

Equally important is designating a media spokesperson early in the process. This will make the process of disseminating public information regarding the investigation more efficient, and prevent conflicting messages from being delivered to the media as a result of multiple contacts.

Safety

For every investigation, the safety of the participants and the public should always be the highest priority. No measurement or sample collection is more important than the health of an individual. The same precautions that apply to all fieldwork should be applied to mollusk kill investigations. Follow proper protocols for entering unknown waters, collecting samples, and using various types of equipment. Equipment should be properly maintained, handled, and cleaned between samples. Appropriate body and foot protection should be worn to protect against general hazards that may occur in water.

Of particular concern are sites that may involve hazardous materials (see Figure 5.1). Do not enter a hazardous spill site until clearance has been given by the agency in charge of spill response. Individuals entering sites with known or suspected hazardous materials must be wearing appropriate body and foot protection. Additional respiratory protection may be needed in the presence of airborne hazards, and additional body, hand, and foot protection may be needed for protection against highly toxic or corrosive chemicals in water or sediments.

Initial Reconnaissance

Following initial notification of appropriate agencies, an initial reconnaissance should begin as soon as possible. An example of a preliminary observation form (Form 5.3) has been provided to assist with this task. Initial information on the magnitude and cause of the kill, any unusual conditions observed, elapsed time since the impact occurred, and whether the kill is still occurring is essential. Valuable information on the cause of the kill is often available in the first few hours of impact. For example, a virtual total kill within a few hours indicates a catastrophic event, such as a chemical spill, whereas a kill that begins slowly, then increases after 1 or 2 d indicates that dissolved oxygen depletion or some type of infectious agent may be involved. These types of observations can determine the types of samples and measurements that need to be collected, as outlined in Table 5.2.

Prompt notification of the suspected responsible party is also important. This action may be needed not only to stop the source of the kill as soon as possible, but will also let them know that an investigation is being initiated. Many times, a potentially responsible party will want to undertake its own investigation of the event. Access to some areas of a water body may require obtaining permission from landowners. Often, the potentially responsible party is also the landowner.

Quick action to stop the source of a kill can be critical. For many types of aquatic contaminants, mussels are able to limit exposure for a period of time, possibly even a few days, by closing their valves. Some snail species may also be able to retract into their shell

Table 5.2. Sampling recommendations for possible mollusk kill causes.

Cause	Typical response	Notable characteristics	Recommended field measurements	Recommended additional samples
Algae bloom	Localized impact; short- or long-term kill	Proliferation of algae; fish kill	pH, clarity, color, dissolved oxygen	Algal density; biological oxygen demand (BOD)
Chemical spill	Impact gradient; short-term kill	Worst near road or industry; mollusks die in place; possible staining of mollusks or substrate; fish kill may precede	pH, clarity, color, dissolved oxygen, conductivity, temperature; gross internal and external exam of mollusks	Parameter list based on suspected source (water, sediment, tissue); common metals if conductivity gradient; biomarkers; sediment and water toxicity tests
Disease	Possibly localized impact; long-term kill	Abnormal color or condition of soft parts; lesions; kill probably confined to mollusks	Gross internal exam of mollusks	Bacteriology, virology, histology
Drought	Short- or long-term kill	Evidence of mollusk movement; stranding; low flow/elevation	Dissolved oxygen, temperature, color, flow/elevation	BOD
Effluent release – acute	Short-term kill; impact gradient	Die in place; discharge near impact; possible staining of shells or substrate; fish kill may precede	pH, clarity, color, dissolved oxygen, conductivity, temperature; gross external exam of mollusks	Parameter list based on suspected source (water, sediment, tissue); common metals if conductivity gradient; biomarkers; sediment and water toxicity tests
Effluent release – chronic	Long-term kill; impact gradient	Evidence of mollusk movement; discharge near impact	pH, clarity, color, dissolved oxygen, conductivity, temperature	Same as above

Table 5.2. Continued.

Cause	Typical response	Notable characteristics	Recommended field measurements	Recommended additional samples
Habitat disruption	Localized impact; short-term kill	Physical disturbance; nearby activity; mollusks stranded or lying on substrate; damage to shells; kill probably confined to mollusks	Clarity, color, total suspended solids (TSS; sediment compaction, color, embeddedness; gross external exam of mollusks	Sediment toxicity tests
Parasites	Possibly localized impact; long-term kill	Parasite infestation may be visible; kill probably confined to mollusks	Gross internal exam of mollusks	Histology
Runoff event	Short-term kill; impact gradient	Nearby agricultural operation; recently salted highway; abandoned minelands or culvert; fish kill may precede	pH, clarity, color, dissolved oxygen, conductivity, temperature, gross internal exam of mollusks	Pesticide analysis; chlorides for road salt; coliform bacteria for livestock wastes; Al, Fe, Mn for abandoned mines; biomarkers; sediment and water toxicity tests
Sediment release	Short-term kill; impact gradient	Noticeable coating of substrate or suspended solids; kill probably confined to mollusks	Clarity, color, TSS; sediment compaction, color, embeddedness	Sediment toxicity tests
Sewage release	Short-term kill; impact gradient	Die in place; treatment plant near worst impact; fish kill may precede	pH, clarity, color, dissolved oxygen, conductivity, temperature; gross internal and external exam	Coliform bacteria, BOD, sediment and water toxicity tests; ammonia
Zebra mussels/ invasive organisms	Possibly localized impact; long-term kill	Proliferation of invasive organism; kill confined to mollusks	Gross external exam of mollusks	Density of invasive organisms

and form a seal with their operculum. When this occurs, fish may be impacted first, such that a quick response in removing the source of the stressor could result in saving mussels from a catastrophic event. However, if mussels maintain valve closure for extended periods, latent mortality may not be observed for several months. Typically, the same agency will be investigating both fish and mollusk kills. Additionally, observations of dying mollusks may provide valuable information concerning possible infectious agents, and to provide real-time dose-response information when associated with concurrent water-quality measurements.

In deepwater habitats, initial observations may not be possible unless divers and equipment are readily available. Although not typically needed, divers are sometimes a necessity and would incur additional costs. An alternative is the use of remote underwater video, which can provide a quick assessment. However, underwater video is of limited use in turbid waters and many mussels may be buried (and not visible) or died in place, rendering video of limited use.

Chemical, Physical, and Biological Measurements

Once the initial information has been gathered, a more intensive process of sample collection and counting dead organisms begins. Hypotheses or conclusive information on the cause of the kill will direct how this process proceeds. Often, more than one cause will be a possibility. The guidelines in Table 5.2 can assist the investigator in selecting the types of chemical, physical, and biological evidence to collect to establish a scientific basis for interpretation of the cause of the kill. To obtain valid data, it is important to maintain proper QA/QC protocols throughout the process of sample collection, handling, preservation, storage, shipping, and analysis.

The most common sampling design will involve establishing a sampling site at the perceived point of greatest impact, with additional sites proceeding along a gradient of impact. At least two reference sites (beyond any observable impact) should be chosen to provide baseline information. On streams and rivers, these should be upstream of the impact. For small streams, if no upstream reference exists, the investigator should establish reference sites on nearby streams as comparable as possible to the impacted stream. For large rivers, the availability of baseline information on the location of dense and diverse mussel assemblages is useful for selecting appropriate sampling sites. If these assemblages are already mapped, then the investigator could randomly select areas within these assemblages to assess for impacts. If such information is not available or is incomplete, the investigator will need to randomly select areas to assess for impacts.

The specimen information form (Form 5.4) and evidence tags (Form 5.5) may be useful during sample collection. A sample identification system should be in place at this time, as well. The specimen information form provides a format for collection of basic measurements on the mollusks. In addition to standard measurements of length/width and age (e.g., estimate external rings), specimens should also be examined for obvious signs of damage or distress. Dead and moribund mollusks with soft parts still attached are the most valuable because several types of analyses can be conducted on the soft parts that will assist in determining the cause of death. If only the shell is available, a determination must be made as to whether it represents a fresh dead mollusk. Under most conditions, a clean, shiny nacre (at the time of collection) is evidence of a recently deceased mollusk (fresh dead) because dulling of the nacre and/or growth of microorganisms on the nacre are typically detectable within a matter of days. Three categories of shell condition are typically used: (1) fresh dead (nacre still lustrous, hinge line still intact, animal probably died within the past few months), (2) weathered dead (nacre chalky, probably dead more than a few months), and (3) subfossil (periostracum missing, shell chalky, probably dead for several decades; Dunn et al. 2000).

Table 5.2 outlines several types of water, sediment, and biological samples that the investigator may wish to collect. For all sample types, reference, impact, and potential recovery sites (if determined) should be sampled to develop comparative information. For some types of samples, an important time element exists regarding either when the sample is collected or when it must be analyzed, as discussed below.

Water samples collected and processed under professional guidelines can provide the most direct measure of the cause of impact; however, they are often only useful when collected while the event is ongoing. Water samples should be collected by a qualified person (such as the state water-quality agency), preserved properly, and documented. Unless the problem is a continuous source of a chronically toxic contaminant, water samples after the event may not detect the problem. Water samples are typically analyzed for metals, organic compounds, chlorides or other anions, fecal coliform bacteria, biochemical oxygen demand (BOD), and algae. Water samples are also obtained for subsequent toxicity tests in the laboratory. Typically, a 1-L sample is sufficient; however, specific needs of analytical procedure should be checked beforehand. For most samples, the collection bottle can be polyethylene. For organic compounds, however, glass bottles with Teflon-lined lids are required. For metals, it is desirable to filter the collected water using a 0.45-μm filter as soon as possible, prior to acidification with nitric acid, so that dissolved forms can be measured. Other preservatives needed are formalin for algae samples and sodium thiosulfate for fecal coliform bacteria. Most other samples require cooling to 4°C. In addition, the investigator should be aware of sample holding times. For example, fecal coliform bacteria samples should be analyzed within 6 h of collection and BOD samples within 48 h.

Acute and chronic toxicity tests may be appropriate. Standard methods for conducting toxicity tests with freshwater mussels are available (ASTM 2013). The laboratory performing the toxicity tests should be consulted on how water samples should be collected and handled.

For sediments, surficial sediments (top 5 cm) obtained with a scoop or grab sample will typically be desired. At each site, a composite of three to five samples from several representative, depositional areas should be collected. In contrast to water samples, sediment samples will often provide residual evidence of contaminants after a kill incident has ceased. Samples for metals and subsequent toxicity tests can be placed in polyethylene bags and cooled to 4°C (not frozen), while samples for organic compounds should be placed in glass jars with Teflon-lined lids and cooled to 4°C. At least 1 L of sediment will be required for each type of sample.

Sediment toxicity tests are most often conducted using the U.S. Environmental Protection Agency's (USEPA 2000b) 10-d *Hyalella azteca* survival and growth test and 10-d *Chironomus tentans* survival and growth test or the American Society for Testing and Materials' (ASTM 2000) *Daphnia magna* or *Ceriodaphnia dubia* three-brood survival and reproduction test, although others may be developed. Sediment-toxicity tests using mollusks can also be conducted and results compared with standard organisms. Sediment-toxicity test standards for mussels are currently being developed. Once these methods are available, sediment-toxicity tests using freshwater mussels would be preferred over other organisms.

Tissue samples are another way of detecting evidence of specific types of impacts after the fact. Tissue levels of contaminants or enzyme biomarkers that respond to specific groups of contaminants can persist for several days following cessation of exposure. Although there is presently no accepted method for correlating a specific contaminant concentration or enzyme level to the death of a mollusk, comparison of these concentrations to levels found in reference organisms can provide evidence that mollusks have been exposed to a particular contaminant or class of contaminants. For tissue samples, at least five specimens should be collected from each sampling site for each type of analysis desired. If possible, specimens should be of the same species and approximate size. The soft parts should be excised from the animal, rinsed with deionized water and blotted dry, and placed in an appropriate container. For organic contaminants, the tissue is typically wrapped in aluminum foil and frozen at less than -20°C until analysis. For metals, each specimen should be placed in an individual polyethylene bag and frozen. For biomarkers, a sample of the tissue of interest should be placed in a cryovial and quickly frozen at less than -80°C until analysis. Useful biomarkers for mollusks include metallothionein as an indicator of elevated metal exposure (Couillard et al. 1993) and cholinesterase as an indicator of exposure to common classes of pesticides (Fleming et al. 1995). Newton and Cope (2007) provide information on a variety of biomarkers depending on the suspected contaminant.

Histological, bacteriological, and virological samples should be collected from moribund (10–15 minutes) animals. Samples can be collected on any persisting tissue, but the time since death will affect whether the disease agents or the agents of decay are being analyzed. Tissue and agent decay is dependent on water temperature, air temperature, flow, and species. The time of tissue collection and time of the event should be recorded and considered in interpretation. Tissue for histological analysis should be placed in vials and immediately fixed with formalin or another suitable fixative. Bacteriological or virological samples should be placed in polyethylene bags and cooled to 4°C (not frozen).

For samples that must be shipped, attention must be given to appropriate packaging to maintain proper preservation, and complete chain-of-custody documentation. A sample chain-of-custody form is provided (Form 5.7). Contact the facility to which you intend to ship the samples and obtain their shipping protocol. Some facilities may limit the days of the week when they will take samples or provide specific samples they prefer based on observations of the animals or the area. The method of shipping and projected time of arrival at the receiving laboratory must accommodate sample holding times. Special labeling or other measures may be required for certain types of samples or preservatives.

General Guidelines for Establishing the Magnitude of a Mollusk Kill

The initial reconnaissance of the site of a mussel kill will, hopefully, establish the extent of the kill zone and where dead mollusks are concentrated. This information, along with previously existing information on the location of mussel assemblages, if available, can be used to determine an appropriate approach to conduct a count of the mollusks. The sampling approach is based on the size of the area to be sampled, the types of habitat that must be sampled, and available manpower, equipment, and time. The time required would be based on sampling conditions and the size of the kill zone.

The ultimate goal in selecting a sampling approach is to maximize the precision of the kill estimate using whatever resources have been committed to the investigation. With impact areas potentially ranging from square meters to square miles in size, there is a similarly wide range in sampling approaches available to fit each situation. For large impact areas, it may be necessary to subsample and extrapolate to the larger area; it is important that any sampling effort be applied in an organized manner. Unless a complete count is being attempted, it is unnecessary and counterproductive to haphazardly count every dead mollusk in sight. In order to produce a valid estimate of the kill, a consistent approach utilizing standardized sampling units and a statistically based sampling design must be followed (Strayer and Smith 2003).

An important consideration is the level of effort that needs to be devoted to the sampling of rare species. Because of the widespread occurrence of endangered mollusks, it is important that good estimates be obtained of any impacts to these species. In general, the more samples that are collected using the most efficient methods, the better the chance of collecting rare species. Several methods have been proposed for estimating the sample size needed to detect rare species (Smith 2006; Tiemann et al. 2009; Huang et al. 2011).

One of the key activities of any investigation of a kill of freshwater mollusks is estimating the magnitude of the impact (i.e., counting dead mussels and snails). Complete counts are rarely possible. Consequently, the investigator must choose an appropriate time and method to sample the area of impact, and use that to estimate the total kill. Dead mussels often stay in place, which means that the same survey methods used to estimate the density of a mussel assemblage of living mussels can be applied to kill assessments. A selection of the most common survey designs is summarized in Table 5.3, modified from Strayer and Smith (2003). State and federal malacologists should also be consulted if possible when determining the study design that is most appropriate for the situation.

Performing a mussel survey to document if a kill has occurred and to document the severity of a given kill is important. However, the design of such a survey and subsequent sampling methods can vary depending on the time of the kill, the location of the kill, the objectives of the survey, sampling conditions, and available resources. Investigators are advised to first contact their state and federal malacologists (if available) to determine if the state in which the kill has occurred in has an accepted standardized protocol for surveying freshwater mussels. If an accepted protocol exists and it meets the objectives of the specific kill, investigators should consider adopting those methods to assess the mussel kill. If a state does not have accepted protocols, surveyors should consult Strayer and Smith (2003)

TABLE 5.3. Survey designs for estimating the magnitude of freshwater mollusk kills.

Survey design	Collection method	Method precision	Time/expense
Complete count	Excavation	Excellent	High per unit area sampled
Systematic (with multiple starts)	Excavation, visual, tactile, dredge, grabs, rake[a]	Good to fair	Moderate to high
Stratified random (habitat strata)	Excavation, visual, tactile, dredge, grabs, rake[a]	Good to fair	Moderate to high
Stratified random (accessibility strata)	Excavation, visual, tactile,, tactile, rake, dredge	Good to fair	Moderate to high
Double sampling intensive)	Exdredge, grabs, rake[a]	Good to fair	Moderate to high
Double sampling (surface/buried)	Excavation, visual, tactile, dredge, grabs, rake[a]	Good to fair	Moderate to high

[a] Excavation is the preferred method, as 30–60% of mussels could be buried. Excavate all substrate within a 0.25 m^2 quadrat down to at least 15 cm, excavate material into an attached bag or container, sieve contents through at least a 6-mm-mesh sieve, sort mussels/snails from debris. Precision and costs will be highest with excavation.

for methods on how to design a survey that will gather the data needed for a specific kill event. Coordination with malacologists in state and federal agencies, as well as the surveys designs outlined in Strayer and Smith (2003), is strongly encouraged. Protocols can be found on the Freshwater Mollusk Conservation Web page (www.molluskconservation.org). The following factors should be considered in the study design:

- Establishing a reference site for comparison of natural mortality (% fresh shells) with mortality at the kill site;
- Sampling as close to the time of the kill as possible, but also sampling several months after the kill to account for latent mortality;
- The time since the kill occurred and resulting impact on the kill count (gaping mussels or shells with tissue are only present for a short time after mussels die; fresh shells without tissue but with lustrous nacre and valves attached, available within a few months of mussel mortality) in the kill count; and
- Using a robust probabilistic design.

It is advisable that kill investigations be approved by an agency malacologist if available before implementation.

The kill enumeration form (Form 5.6), or a similar form, will be needed to provide a record of the count of dead mollusks to estimate the total kill. This form includes a description of the sampling unit, with spaces for entering the count of each species collected from the unit. It may be useful to keep separate counts of adults and juveniles because of the known difficulty in collecting juveniles. Separate counts will allow separate estimates to be made of total kills of adults and juveniles. Better precision can be expected for the adult kill estimate compared to either the juvenile estimate or an estimate made for samples where adults and juveniles were not distinguished. Typically, mussels less than 5 years old are considered juveniles. Aging by counting of annuli is acceptable for young mussels, recognizing that this method tends to underestimate the actual age of a mussel, particularly for older specimens (Neves and Moyer 1988).

The enumeration form also distinguishes among live animals, fresh dead, and weathered shells. The most difficult of these to define are fresh dead mussels, particularly when the soft tissues are gone. If soft tissues are present, live animals should contract their tissues and close their valves in response to a probe. If there is no response, the animal can be considered dead. There are many situations, however, when the soft tissues are lost (or scavenged) within a short time following death of the animal. For collected shells, as defined in this publication, those possessing clean, shiny nacres should be considered fresh dead. If the number of fresh dead mollusks is too numerous to enumerate, a correction factor can be developed by comparing ratios of shells classified as fresh dead at impacted sites to those at reference sites. The investigator may also consider a fourth category for mussels in obvious distress, as evidenced by widely gaped valves or a weak valve-closure response, but not yet dead. Finally, the investigator should report the fate of each counted mollusk (i.e., returned to the water, saved and tagged as evidence, etc.).

Counts should not include invasive mollusks such as Asian clams, Zebra mussels, Quagga mussels, and Chinese mystery snails *Cipangopaludina chinensis* unless these animals are of interest as a potential cause of the kill.

Record Keeping

It is important that detailed records be kept of the entire investigation. These include dates and details of important events, persons contacted, sampling efforts, counts of mollusks, disposition of specimens, and other details. This information will be needed to prepare a final report of the investigation, but will also be essential for possible legal proceedings and to document investigation expenses.

Key records that will assist in the record-keeping process include a chronological log of events, a sample identification system, chain-of-custody forms, expense receipts, and timesheets. In addition, concisely labeled photographs and maps are valuable. Examples of useful photographic evidence include contaminant sources or other possible causes of the kill, unusual environmental conditions, the magnitude of the kill and species affected, and any corrective actions taken.

Public Relations

Kills are newsworthy, especially if they are highly visible to the public in the form of masses of dead fish floating on the water or washed up on shore. As mentioned previously, a single media spokesperson from the lead investigating agency should be appointed. This person should be available for questions from the public and media, and should be the source of any news releases. Information released to the public should be carefully considered. Do not prematurely assign blame for a kill incident, and provide only factual accounts of the investigation. Examples of appropriate information include the extent and duration of the kill, agencies involved in the investigation, and any public safety concerns associated with the kill.

Good public relations also includes dealing with local residents in the vicinity of the kill incident. These individuals may have a proprietary concern for the resource. Some may also be directly impacted by activity associated with the investigation, such as site access. Keeping local residents informed during the process can make the investigation progress go smoothly.

Reporting

Preparation of a final report of the mollusk kill investigation is typically the responsibility of the lead agency. It is critical that sufficient effort be applied to the report so that it will withstand anticipated legal challenges. This effort will require a thorough and accurate accounting of the investigation and a careful interpretation of collected data and include, if appropriate, maps, photographs, and analytical reports from outside laboratories or specialists. It is important to stick to the facts and scientific interpretation of those facts. Do not engage in editorializing, emotional statements, or personal opinions. Once the final report is completed and has undergone internal review, a summary will be need to be prepared for use as a news release.

An example of a final report is provided as Form 5.8. The report provides information on the initial reporting and reconnaissance of the kill incident, where and when it occurred, and who was involved in the investigation. Collected chemical, physical, and biological data are summarized, including samples and specimens collected for laboratory analysis. The results of the data collection and analysis are discussed, and a cause for the kill is proposed along with supporting evidence. Following this discussion, recommendations for corrective action can be included. These recommendations should address questions such as

Were the mollusk assemblages of the water body significantly damaged by the kill?
Should there be replacement by stocking?
What action should be taken to prevent future recurrences?
Should a claim for damages be filed against the responsible party?

The final sections of the report detail calculations of a total loss estimate of mollusks and of investigation expenses. Appendix F provides hatchery production costs of freshwater mussel species.

Endangered Species Considerations

If the kill occurs in an area known to contain federally endangered mollusks or if a specimen (live or dead) of an endangered mollusk is identified during the investigation, the U.S. Fish and Wildlife Service must be immediately involved in the investigation. Because of the significant percentage of the mollusk fauna that is currently listed as threatened or endangered by the federal and/or state governments, with many more candidate species, there is a good chance that an impact to a water body containing a diverse assemblage of mollusks will involve at least one endangered species (see Appendix E).

The presence or suspected presence of a state or federal endangered or threatened species has implications for how a kill assessment is carried out. If a complete count cannot be made due to the size of the kill, accessibility, or other factors, then the sampling approach should be designed to provide as good an estimate as practical of losses of endangered species. This typically entails the use of sampling methods that sample the largest number of sampling units most efficiently (e.g., visual searches by wading, snorkeling, or diving).

Chapter 6

Costs and Economic Valuation of Freshwater Mussel Kills

Statement of Applicability

The total monetary damage from a mussel kill is made up of several primary components: ecological (ecosystem services), use (habitat for fish and other benthic organisms), and nonuse values (aesthetic appreciation), plus replacement costs (Southwick and Loftus 2003). Ecological, use, and nonuse values should be included as part of monetary damages whenever they can be reliably quantified. However, because information to accurately estimate ecological, use, and nonuse values for mussels are only occasionally available, the process presented in this chapter is based on using replacement costs as a *conservative* method to determine restitution for killed mussels.

The *true* costs associated with a mussel kill cannot be fully accounted for using replacement costs alone. Hatchery-raised mussels may not be equivalent (in terms of survivorship, genetic make-up, or the ecological services they provide) to native species killed, and habitat restoration may be needed to ensure the site can host viable mussel populations.

Until research can be conducted into the full range of values provided by mussels, the replacement costs presented in this chapter must be considered a conservative and minimum method of assigning restitution after mussel kills. *Production cost data or life history data local to the mussel kill site, region, genera or species should be used in preference to the general economic data provided in this book.*

These estimated replacement costs are not applicable when time and resources are available to develop more current and relevant estimates through the Natural Resource Damage Assessment (NRDA) process.

These estimated replacement costs do not account for species that require life history research before propagation can be successful, and/or species that may be considered exceptionally difficult to raise. Additional damages are warranted for such species because the costs listed in this document are averages and likely underestimate the true cost of replacing hard to propagate species. Likewise, if data exist demonstrating the mussel population impacted by a kill has unique traits uncommon in other areas, additional damages are warranted

Replacement cost data should not be used for species that are currently listed as threatened, endangered, or otherwise protected where specific laws and regulations regarding such species may specify civil or criminal penalties. To ensure that full compensatory damages and restoration activities are realized, federal and state agencies should be consulted regarding kills involving federally protected (or state protected) species. In addition to federal claims, many states have threatened and endangered lists and related civil and criminal penalties that may take precedence over the guidelines presented in this document. In some instances, states may be able to claim restitution damages in addition to federal claims.

After some kill events, it may not be possible to restore the site due to irreparable habitat damage. In these instances, authorities should defer to federal NRDA guidelines.

In addition, if the habitat of a kill site is marginal and restocked mussels are not expected to survive or successfully reproduce, replacement and restoration efforts should be directed to locations where stocked mussels are more likely to survive and reproduce.

Replacement Costs

The replacement costs of mussels are composed of several elements:

1. Production costs—the costs associated with producing juvenile mussels in a propagation facility to an age/size described later in this chapter, including the search for broodstock (gravid females) and any research, collection, rearing, and maintenance costs related to host-fish species;
2. Restocking costs—the costs associated with restocking the kill site over one or multiple years and restoring habitats; and
3. Investigation, monitoring, and administration costs—the costs of investigating, monitoring, and managing the kill.

Production Costs

The mussel production costs in this chapter are based on costs reported by mussel propagation facilities in 2017 (see Appendix F). Production costs for a particular species can vary regionally based on several factors, such as the acquisition of broodstock females and degree of endemism of a population. When possible, after a mussel kill, cost data from any local propagation facilities raising and stocking the affected species should be used in lieu of the data herein. Otherwise, the cost data provided in this chapter may be applied as proxy values. In some cases, freshwater mussels might not be raised in facilities to adulthood before being released. Some mussel propagation facilities release mussels as juveniles. In these cases, two additional issues must be addressed for restoration of the mussel population:

1. Juvenile mortality—When stocked as juveniles, time is needed for mussels to reach an age similar to those that died in the kill. The mortality incurred between the time of stocking and the time that the mussels reach a particular age must be factored into the calculation of replacement cost. The methodology in this book is based on deter-

mining the cost of restocking a kill site with the number of taggable-sized mussels[1] that must be stocked to replace each mussel of each age that died during the kill. In situations where mature mussels are being stocked, these adjustments are not appropriate.
2. Age to sexual maturity—An adult mussel is defined as one that has reached sexual maturity. After most mussel kills, it is often not possible or feasible to estimate the age of all mussels, particularly the youngest and oldest mussels. Therefore, despite the fact that many adult mussels that died during a kill may have been reproductively more prolific than younger replacement mussels, a conservative recommendation is made to assume that the adult component of the mussel population is restored when the replacement mussels reach sexual maturity.

A goal of any restocking effort should be to re-establish a like-sized mussel population composed of species present before the kill that has the ability to regenerate itself through natural reproduction. This typically will require multiple stockings over a period of several years to establish multiple age-classes, which is the recommended stocking procedure. However, unless reliable scientific data becomes available as to when a restocked mussel population is self-sustaining, defer to the recommendation that mussels are considered restored when all the replacement mussels (all age-classes) have reached sexual maturity.

Restocking Costs

If mussels are actually restocked after a kill, the cost of transporting mussels from the propagation facility to the stocking site and the associated costs of vehicles, fuel, water, personnel, and overhead should be included in total replacement costs. If the cause of the mussel kill also caused damage to habitats or limited the viability of a sustainable population of host fish, the costs of chemical and physical rehabilitation—supplies, equipment, personnel, overhead, and replacing lost host fish species—should be added.

Investigation, Monitoring, and Administration Costs

Investigating a kill, estimating the magnitude of loss, dealing with the public, and coordinating all the activities engendered by a kill are labor-intensive. Relevant personnel costs—which may be incurred by several agencies—should be recovered, as well as the costs of transportation, field equipment, overhead, state permitting, reporting, and other relevant costs. Such costs include the time and expense to monitor long-term effects of the kill and the effectiveness of restocking efforts.

Source of Production Cost Data

Replacement Mussels

The production costs provided here are derived from a 2017 survey of mussel propagation facilities in the United States. This survey provided information regarding the costs as-

[1] Taggable size generally refers to a mussel, typically greater than 15mm, capable of supporting an external physical tag.

sociated with producing juvenile mussels, the expected costs for mussels not currently in production, and the costs of raising large numbers of mussels to replace those that died in kill events. Unlike the previous 2003 edition of this publication, costs are provided for each genus, not per species, given the limited availability of production data for many species. The production costs provided in this publication report the weighted average costs of producing each genus in large quantities sufficient for restocking. Weighting was based on the number of each genus produced within each facility. Note that the *production* costs contained in this publication do *not* include investigation/administration or complete restocking costs, as outlined earlier. These costs should be quantified and included as monetary damages on a case-by-case basis, as the subsequent examples will illustrate.

Juvenile Mussels

The survey respondents also provided the number of each mussel species needed to be stocked for 1,000 mussels to survive from taggable size to 5 years of age. Because taggable size and survival vary with species, geographic range, and other factors, the rates varied based on a range of taggable sizes (2 months to 4 years) surviving to 5 years old. Thus, a mean survival rate for all species was not calculated. If data regarding survival rates specific to the species (or population) that died in the kill and the age at sexual maturity are known, these data should be used in lieu of any generalized survival rates. Investigators should consult with a malacologist and propagation facility, if possible, for ages to sexual maturity, survival rates, and recommended stocking sizes and ages for purposes of calculating the numbers needed for stocking each species in a given locality. Investigators also should consider that annual survival is related to life span and is nonlinear. Survival may increase dramatically after year one and may approach an asymptote around age 3 or 4 (W. R. Haag, U.S. Forest Service, personal communication; see also Haag 2012). If no species-specific information is available, the best expert opinion (Freshwater Mollusk Conservation Society, personal communication) suggests that mean survival from stocking at 2 months of age to 5 years of age approximates 15%, but this could vary and should be updated as more specific information becomes available. This rate is derived from mean survival rate of 11 mussel species from 2 to 5 months of age to the following year (mean 0.6, range 0.34 to 0.98, Haag 2012). The mean rate of 0.6 was multiplied by 5 years to obtain a rate of 30% from 2 months to 5 years of age and was then multiplied by 50% to account for the fraction that would be nondetectable, as juvenile mussels are hard to detect without considerable effort.

If at all possible, when evaluating a mussel kill, the age composition of mussels that have been impacted should be determined. If this is not possible, all mussels killed should be assumed to be mature mussels. From the age composition and survival rate of mussels to 5 years, estimates can be made of the number of juvenile mussels that need to be stocked in order to replace those of each age-group that were impacted. For adults (or ≥5 years if maturity is not known), this is done by dividing the estimated number of mature mussels that died in the kill by the known percentage surviving to maturity or the average percentage calculated from the survey results. An estimate of survival of each age-class can be generated with the general exponential formula

$$N_t = N_0 e^{-rt}$$

or, rearranging to determine the number that need to be stocked at time 0,

$$N_0 = N_t / e^{-rt},$$

where

N_t = number of mussels surviving to time t after stocking
N_0 = number of mussels originally stocked
e = the base of natural logarithm
r = instantaneous rate of decline (calculated from annual survival/mortality)
t = elapsed time since stocking (i.e., age), in years

Transportation Cost

If the actual cost of transporting replacement mussels to the stocking site is not known, an estimate of $1.42 per mile can be used unless there is more specific localized or updated data. This is the average cost reported by fish hatcheries to deliver and stock fish (see Chapter 4). Note that this does not include personnel, preparation and loading of transport vehicles or related costs at the propagation facility. Transportation costs incurred to collect broodstock, or gravid females, should also be recovered, including lodging, per diem, and other travel-related expenses.

Adjusting for Inflation

Due to inflation, costs incurred by mussel production facilities will change over time. The prices, or replacement costs, provided in this text are based on costs reported by mussel producers in 2017. It would be reasonable for mussel kill investigators to adjust these 2017 prices to reflect the year in which the kill occurred using the producer price index (PPI). The PPI "measures the average change over time in the selling prices received by domestic producers for their output" (U.S. Bureau of Labor Statistics 2002). Current PPI information can be obtained from the U.S. Department of Commerce, Bureau of Labor Statistics (www.bls.gov/PPI/).[2] The producer price index for all commodities (PPIACO) has been applied in this publication. See Chapter 4 for an example of the application of PPI.

Present Value

If restitution payments for mussel damages are made today for restoration activities to occur later (such as stocking of mussels in future years to establish age cohorts in the restored population), it is appropriate to discount those payments.

Rather than relying on natural reproduction to repair damaged mussel populations, mussel populations should be restocked over a period of years whenever possible. In this case, it would also be appropriate to recover the costs of restocking over time in one up-front payment. Suppose an equal number of mussels will be restocked immediately in the current year and at the beginning of the next 3 years. Also, assume the costs of restock-

[2] Links are current at the time of this publication but may change in time.

ing (including propagation facility and transportation costs) are $25,000 per stocking per year. To calculate the present value of these multiple years of restocking, the general EV_{pv} formula discussed in Chapter 4 is applied as follows:

$$PV = EV_{pv} = EV_0 + \frac{EV_1}{(1+r)} + \frac{EV_2}{(1+r)^2} + \frac{EV_N}{(1+r)^N},$$

where

EV_{pv} = present value of future stream of lost economic values
EV_0 = economic value in the present year;
EV_1, EV_2, EV_N = economic values in the first, second, and Nth years after the present year;
r = interest (discount) rate

$$EV_{pv} = \$25{,}000 + \frac{\$25{,}000}{(1+0.05)} + \frac{\$25{,}000}{(1.05)^2} + \frac{\$25{,}000}{(1.05)^3},$$
$$= \$25{,}000 + \$23{,}809 + \$22{,}676 + \$21{,}596$$
$$= \$93{,}081 = \text{the value in today's dollars of the costs of stocking mussels over the next 4 years.}$$

Note that the $25,000 cost of restocking immediately in the current year does not have to be discounted because it is already in present value or current dollar terms.

The above example is simplified for demonstration purposes. In addition to propagation facility and transportation costs, it would also be appropriate to include all investigation, administration, site restoration, and monitoring costs expected during the rehabilitation period.

Limitations and Exceptions When Applying Replacement Costs

1. Any production cost or survival rate data specific to the mussel kill site, region, genus, or species should be used in preference to the general economic data provided in this book. Otherwise, the production cost data provided in this document may be applied as proxy values. The methodologies described here still will be applicable.
2. Production cost data (Appendix F) should not be used for species that are currently listed as threatened, endangered, or otherwise protected at the federal level. Production cost concepts do not apply to such species. Further, state and federal laws and regulations regarding such species may specify civil or criminal penalties. Mussel kills involving federally protected species typically require extensive interagency coordination between state and federal agencies.[3] Many states have threatened and endan-

[3] A state claim submitted without federal trustee input has the potential to limit the federal trustee's ability to fully assess injury, pursue damages, and implement site-specific restoration activities due to double-recovery prohibitions in federal regulations. Federal entities are required by policy to consult with state co-trustees before proceeding with formal damage assessments in state waters.

gered lists and related civil and criminal penalties that may take precedence over the guidelines presented in this publication. A list of federally threatened and endangered mussels as of July 31, 2017 is found in Appendix E.

3. For mussels that are protected by special state statutes or policies (e.g., listed on the state threatened and endangered list) or for mussels that may be considered exceptionally difficult to propagate due to life history characteristics, consult with the state and local experts to confirm taxonomic status of the species (and any special laws or statutes that may apply) before applying the values in Appendix F. It may be appropriate to apply additional monetary damages above and beyond the production costs listed in this document. These additional damages would be warranted because the costs listed in this publication are averages and likely underestimate the true cost of replacing hard-to-produce species.
4. After some kill events, restoration of the site may not be possible due to irreparable habitat damage. In these instances, authorities may defer to federal guidelines.
5. In addition, it is recommended that replacement and rehabilitation efforts be directed to locations where restocked mussels are more likely to survive and successfully reproduce if the habitat of a kill site is marginal and restocked mussels are not expected to either survive or successfully reproduce. Consultation with state and federal partners in that region is important to determine these appropriate locations.

Estimating Replacement Costs

Evaluating the monetary damages from a mussel kill should begin as soon as the kill is reported. The extent of the kill and the number of mussels that died needs to be determined (see Chapter 5). The methods employed to quantify damages vary with the characteristics of a mussel kill and with the biological and economic information available. Methodologies are presented in this section to cover three mussel kill scenarios. **These scenarios are simplified for purposes of exposition, and they do not cover all possibilities.** In the aggregate, however, they indicate procedures that can be used—in some appropriate combination—for any mussel kill. The procedures do not identify all potential economic losses resulting from mussel kills; for example, they do not measure ecological use (ecosystem services, habitat for fish and other benthic organisms, etc.), and nonuse values (aesthetic appreciation). In deciding which scenario to apply, the management agency must decide whether to restock. Then, the manager can identify the scenario below that best fits the situation and follow the steps given.

Compared to fish, which may repopulate a kill site relatively quickly, mussels require long periods of time, often decades, to repopulate. In many cases, mussel populations have been unable to recover. The primary value of mussels comes from their numerous documented ecosystem services, which, to date, have only been partly quantified. It is strongly recommended that mussels be replaced after a kill as the true long-term damages to the ecosystem may be significant. Even though replacement costs will likely underestimate the long-term ecological and use values that were diminished in response to the mussel kill, replacement cost methods provide alternative, conservative, restitution amounts.

Based on the procedure of stocking numerous juvenile mussels to replace one killed adult mussel, there is no need to seek compensation for the time required for juveniles to grow to maturity. The numerous stocked juvenile mussels are assumed to replace the ecological services once provided by the adult that died. While there may be errors in this assumption, this assumption is recommended until better scientific data become available.

Follow the steps below to determine the proper restitution for differing mussel kill scenarios:

Scenario 1: Natural Production Will Be the Means for the Resource to Repair Itself

Under this scenario, no stocking is conducted. Rather, repopulation of the damaged mussel populations is achieved by natural reproduction by the mussels that remain after the kill. In this case, the costs of the mussels, if they were to be produced and stocked, is used as a surrogate value to assign damages for the responsible party[4] (i.e., the calculated costs are used to assess penalties, even though no stocking occurs).

A. Determine the investigation, adistration, rehabilitation, and monitoring costs.
B. Using the formula presented earlier in this chapter, estimate the number of mussels of taggable size that would be required if restocking were to occur to replace those that died in the kill.
C. For each species, multiply the number of mussels estimated in step B by its relevant production cost as listed in Appendix F. Please note that costs are reported by genus.
D. Determine the cost of conducting this economic analysis, including contractors and staff time, not included in any of the steps listed above.
E. Add the costs (steps A + C + D) to arrive at the total amount of the penalty to be assessed.

Scenario 2: Mussels Will Be Translocated from Another Location

Under this scenario, mussels of the same species as those killed are collected from another viable population and translocated to the damage site. Before restocking occurs, due consideration must be given to assessing the genetic implications of introducing mussels from another location, the impact of removing mussels from the population from which they are harvested, and the likelihood that the translocated mussels will survive and reproduce in the new location. This scenario is used when mussels are not available from any propagation facility. Translocating mussels from another site may re-establish mussel populations faster than stocking juvenile mussels.

Restitution for this scenario typically will take place after mussels have been translocated. To determine the restitution amount using this option, carefully document all costs associated with all translocation activities:

A. Determine the investigation, administration, travel, and rehabilitation costs, and estimate monitoring costs. This includes costs associated with both the kill site and the site(s) from which mussels will be collected for translocation.

---

[4] This process is solely based on the replacement costs of mussels. It does not specifically measure ecological use or nonuse values of mussels. Such values should be included if they can be quantified.

B. Document all labor costs (both staff and contracted labor) and expenses required to find, collect, and translocate mussels, as well as monitoring of the translocated mussels. This includes transportation costs, management, overhead, permitting, and other costs that can be directly related to investigating the kill and translocating the mussels.
C. Calculate the replacement costs for the mussels removed from the collection site, as described in this chapter.
D. Determine the cost of this economic analysis, including contractors and staff costs not included in any of the steps listed above.
E. Add steps A, B, C, and D to arrive at the final restitution value.

Scenario 3: Restocking with Immature Mussels

Under this scenario, taggable-sized mussels are stocked from a production facility to reestablish a mussel population (or portion of a population) that died in a kill. Procedures for calculating the monetary damages are similar to Scenario 1, except that efforts are made to produce and restock mussels.[5]

A. Determine the investigation, administration, rehabilitation, and monitoring costs.
B. Using the formula presented earlier in this chapter, estimate the number of mussels required to replace those that died in the kill.
C. Determine how many juvenile mussels need to be stocked each year over the next N years to reintroduce age cohorts within the mussel population. Using the costs presented in Appendix F, determine the production costs of the mussels needed to be stocked each year.
D. For each year when mussels are to be stocked (step C), add to the annual mussel production costs any additional recurring costs such as labor and management, facilities, and other expenses directly related to stocking the kill site. Only add costs not already included in step A.
E. Determine the discounted value for the results of step D using the discounting, or present value, techniques described earlier in this chapter.
F. Determine the cost of this economic analysis, including contractors and staff time, not included in any of the steps listed above.
G. Add steps A + E + F to arrive at the final restitution value.

Examples

Mussel kill valuation procedures for the three scenarios of the previous section can be illustrated with a simple hypothetical example. An accidental chemical spill occurred in a stream with a viable population of *Lampsilis cardium*. No permanent damages occurred to the habitat, but investigators have estimated that 15,000 mussels were killed. For illustra-

[5] Often the number that can be produced is less than the number killed: the replacement costs should be based on the number required to replace the number that was killed and not on limitations of production capabilities. Also, note that additional production costs may need to be included to cover research and development for species that are difficult to raise.

tive purposes only, assume that *L. cardium* reaches taggable size in this stream at 6 months and has a survival rate of 10% from 6 months to 5 years.

Scenario 1: Natural Production Will Be the Means for the Resource to Repair Itself

A. The cost of the investigation, administration, and monitoring is estimated at $64,000.
B. From the field investigation, estimates of the total number and age composition of mussels killed are

<age 2:	4,000
Age 2:	3,000
Age 3:	2,000
Age 4:	1,500
Age 5+:	4,500
TOTAL:	15,000

(If ages cannot be estimated, then assume all are adult [mature] mussels.)

Using the formula presented earlier, we determine the number of mussels per age-class that need to be stocked to replicate the number killed in the chemical spill. Using the age to maturity and survival to maturity (converted to an instantaneous rate of 0.51), above, we calculate[6]

Age 0–2:	$N_0 = 4{,}000/e^{-0.51}$	= 6,661 mussels
Age 2:	$N_0 = 3{,}000/e^{-0.51(2)}$	= 8,320 mussels
Age 3:	$N_0 = 2{,}000/e^{-0.51(3)}$	= 9,236 mussels
Age 4:	$N_0 = 1{,}500/e^{-0.51(4)}$	= 11,536 mussels
Age 5+:	4,500/0.10	= 45,000 (adult mussels)

80,753 = the estimated total number of mussels needed to replace those killed.

Note that the number of mussels in the Age 5+ category does not account for any mortality after age 5 and therefore is likely an underestimate of the true number of mussels needed to exactly replace those killed but nonetheless provides a good indication of the number needed to reach a minimum age of 5 years.

C. Using Appendix F, the estimated cost associated with producing a taggable-sized *Lampsilis* is $27.20. With 80,753 mussels at $27.20 each, the estimated production cost is $2,196,482.
D. The cost to conduct this economic analysis was $300.
E. The total restitution amount (steps A + C + D) is $2,260,782.

Scenario 2: Mussels Will Be Translocated from Another Location

A. Investigation, administration, and monitoring costs were estimated to be $82,000 ($64,000 for the kill site and $18,000 for the site where mussels will be taken for translocating).

[6] For an explanation of the relationship between instantaneos survival and annual survival, consult, for example, Ricker (1975, chapter 1 and appendix 1).

B. Labor costs associated with collecting and translocating the mussels are $10,000.
C. The replacement costs are calculated using the same steps presented in Scenario 1, steps B and C, which equal $2,196,482.
D. The cost of this analysis was $300.
E. The final restitution value is $2,278,782.

Scenario 3: Restocking with Taggable-Sized Mussels

A. The cost of the investigation, administration, and monitoring is estimated to be $64,000.
B. Investigators determine the size structure of the mussels killed is

<age 2:	4,000
Age 2:	3,000
Age 3:	2,000
Age 4:	1,500
Age 5+:	4,500
TOTAL:	15,000

Using the formula presented earlier, we estimate the number of mussels per age-class that need to be stocked to replace the number killed in the chemical spill. Since we do not know the exact age to maturity or survival to maturity, we use the values of 5 years and 10% (converted to an instantaneous rate of 0.51), respectively:

Age 0–2:	$N_0 = 4{,}000/e^{-0.51}$	= 6,661 mussels
Age 2:	$N_0 = 3{,}000/e^{-0.51(2)}$	= 8,320 mussels
Age 3:	$N_0 = 2{,}000/e^{-0.51(3)}$	= 9,236 mussels
Age 4:	$N_0 = 1{,}500/e^{-0.51(4)}$	= 11,536 mussels
Age 5+:	4,500/0.10	= 45,000 (adult mussels)

80,753 = the estimated total number of mussels needed to replace those killed.

C. The following stocking schedule is determined as the best opportunity to restore the age cohorts identified in the kill:

Year 1: stock 45,000 mussels to replace the age 5+ that were killed
Year 2: stock 11,536 mussels to replace the age 4 that were killed
Year 3: stock 9,236 mussels to replace the age 3 that were killed
Year 4: stock 8,320 mussels to replace the age 2 that were killed
Year 5: stock 6,661 to replace the age 0–2 that were killed

Using Appendix F, the cost associated with *Lampsilis* is $27.20 each and the cost of replacement mussels for each year will be

Year 1: 45,000 mussels @ $27.20 each = $1,224,000
Year 2: 11,536 mussels @ $27.20 each = $313,779
Year 3: 9,236 mussels @ $27.20 each = $251,219
Year 4: 8,320 mussels @ $27.20 each = $226,304
Year 5: 6,661 mussels @ $27.20 each = $181,179

D. It is determined that the agency cost to stock the kill site each year is $2,000 (expressed in today's dollars), so, total costs by year are

   Year 1: $1,224,000 + $2,000 = $1,226,000
   Year 2: $313,779 + $2,000 = $315,779
   Year 3: $251,219 + $2,000 = $253,219
   Year 4: $226,304 + $2,000 = $228,304
   Year 5: $181,179 + $2,000 = $183,179

E. Calculate the present value or discount value of the replacement mussels. Assume that the interest rate charged by your agency is 2.1:

$$EV_{pv} = \$1{,}226{,}000 + \frac{\$315{,}779}{(1+0.021)} + \frac{\$253{,}219}{(1+0.021)^2} + \frac{\$228{,}304}{(1+0.021)^3} + \frac{\$183{,}179}{(1+0.021)^4}$$

$$= \$1{,}226{,}000 + \$309{,}284 + \$242{,}910 + \$214{,}504 + \$168{,}567$$

$$= \$2{,}161{,}266$$

F. The cost to conduct this economic analysis was $300.

G. The total restitution amount (steps A + E + F) is $2,225,566

It is apparent from these examples that the way in which a mussel kill is valuated can result in differing costs. The best approach should be the one that reflects the most scientifically sound data that are available for restoring the resource regardless of cost.

Consideration of Mussel Values

In the economic literature, the total economic value of a natural resource includes values involving active use in the present or future (use values) and values that are independent of present or future use (nonuse values). The economic and ecology literature also recognize broad values of natural resources as components and contributors to healthy ecosystems and ecosystem services (ecological values). In this section, use, nonuse, and ecological values of mussels are described in more detail.

Human Use Value

Use (or user) value of mussels is the value of the resource to people who use mussels but whose use has been curtailed by the kill. It is estimated from the time and money people actually or potentially spend to use a mussel resource or the profit they derive from its use. If such values can be quantified (such as profits from commercial harvesting, use in bait, etc.), these values should be included because they represent losses resulting from the kill. Use values may also include latent, future-use values, such as opportunities that mussels may provide in the future but are currently unavailable now. Use values, alone, will not account for all losses resulting from a mussel kill. Many species are not used by humans but still provide important ecological services.

Human Nonuse Value

Conceptually, nonuse values of mussels include option, bequest, and existence values. The option value of mussels is the value of retaining the option to use mussels in the future, whether or not this option is ever exercised. The bequest value of mussels includes the value an individual places on providing mussels for other people in the same generation (intragenerational bequest value) and for people in future generations (intergenerational bequest value). Existence value is the value an individual places on knowing that mussels exist, independent of any current or future active use of mussels. Although the concept of nonuse valuation is documented throughout the economic literature, nonuse values for mussels have not been empirically estimated. Nonuse values are thought to be slight for mussel kills of small or modest size (the magnitudes for which these guidelines are most appropriate). For lack of data, nonuse values will not be considered further in this book. However, if and when such information becomes available, their inclusion should be considered.

Ecological Value

Mussels provide numerous ecological benefits (Vaughn and Hakenkamp 2001; Vaughn 2010, 2017), and their greatest contributions may come from these ecological roles, including their value as water purifiers. Loss of mussels may result in degradation of water quality, leading to negative impacts on the overall health and productivity of the ecosystem, including human recreational and commercial uses. Other ecological benefits include providing food for wildlife, nutrient recycling, streambed stabilization, fish spawning habitat, and habitat structure for invertebrates. These values, though not quantified at the time of this publication's release, may be substantial.

Note that legally protected critical habitats that include mussels need special consideration when determining restitution after a damage event has occurred, just like protected species that inhabit them. Federal natural resource damage assessment guidelines or specific state guidelines may take precedence in such situations.

References

AFS (American Fisheries Society). 1982. Monetary values of freshwater fish and fish-kill counting guidelines. AFS, Special Publication 13, Bethesda, Maryland.

AFS (American Fisheries Society). 1992. Investigation and valuation of fish kills. AFS, Special Publication 24, Bethesda, Maryland.

AFS (American Fisheries Society). 1993. Sourcebook for investigation and valuation of fish kills. American Fisheries Society, Special Publication 24 supplement, Bethesda, Maryland.

APHA (American Public Health Association), American Water Works Association, and Water Environment Federation. 2012. Standard methods for the examination of water and wastewater, 22nd edition. APHA, Washington, D.C.

ASTM (American Society for Testing and Materials). 2000. Test method for measuring the toxicity of sediment-associated contaminants with fresh water invertebrates. ASTM International, ASTM E1706-00, West Conshohocken, Pennsylvania.

ASTM (American Society for Testing and Materials). 2013. Standard guide for conducting laboratory toxicity tests with freshwater mussels. ASTM International, ASTM E2455-06, West Conshohocken, Pennsylvania.

Anthony, J. L., and J. A. Downing. 2001. Exploitation trajectory of a declining fauna: a century of freshwater mussel fisheries in North America. Canadian Journal of Fisheries and Aquatic Sciences 58:2071–2090.

Axon, J. R., L. Hart, and V. Nash. 1980. Recovery of tagged fish during the Crooked Creek Bay rotenone study at Barkley Lake, Kentucky. Proceedings of the Annual Conference Southeastern Association of Fish and Wildlife Agencies 33(1979):680–687.

Ball, R. C. 1948. Recovery of marked fish following a second poisoning of the population in Ford Lake, Michigan. Transactions of the American Fisheries Society 75:36–42.

Brown, C. J. D., and R. C. Ball. 1943a. An experiment in the use of derris root (rotenone) on the fish and fish-food organisms of Third Sister Lake. Transactions of the American Fisheries Society 72:267–284.

Brown, C. J. D., and R. C. Ball. 1943b. A fish population study of Third Sister Lake. Transactions of the American Fisheries Society 72:177–186.

Carlander, K. D. 1969. Handbook of freshwater fishery biology, volume 1. Iowa State University Press, Ames.

Carlander, K. D. 1977. Handbook of freshwater fishery biology, volume 2. Iowa State University Press, Ames.

Carson, R. T. 2012a. Contingent valuation: a comprehensive bibliography and history. Edward Elgar Publishing, Cheltenham, UK.

Carson, R. T. 2012b. Contingent valuation: a practical alternative when prices aren't available. Journal of Economic Perspectives. 26(4):27–42.

Couillard, Y., P. G. C. Campbell, and A. Tessier. 1993. Response of metallothionein concentrations in a freshwater bivalve (*Anodonta grandis*) along an environmental cadmium gradient. Limnology and Oceanography 38:299–313.

Davies, W. D., and W. L. Shelton. 1983. Sampling with toxicants. Pages 199–213 *in* L. A. Nielsen and D. L. Johnson, editors. Fisheries techniques. American Fisheries Society, Bethesda, Maryland.

Dunn, H. L., B. E. Sietman, and D. E. Kelner. 2000. Evaluation of recent unionid (Bivalvia) relocations and suggestions for future relocations and reintroductions. Pages 169–183 *in* R. A. Tankersley, D. I. Warmolts, G. T. Watters, B. J. Armitage, P. D. Johnson, and R. S. Butler, editors. Freshwater mollusk symposia proceedings. Ohio Biological Survey, Special Publication, Columbus.

Fleming, W. J., T. P. Augspurger, and J. A. Alderman. 1995. Freshwater mussel die-off attributed to anticholinesterase poisoning. Environmental Toxicology and Chemistry 14:877–879.

Haag, W. R., and J. D. Williams. 2014. Biodiversity on the brink: an assessment of conservation strategies for North American freshwater mussels. Hydrobiologia 735:45–60.

Hayne, D. W., G. F. Hall, and H. M. Nichols. 1967. An evaluation of cove sampling of fish populations in Douglas Reservoir, Tennessee. Pages 244–297 *in* Reservoir fishery resources symposium. American Fisheries Society, Southern Division, Reservoir Committee, Bethesda, Maryland.

Hayne, D. W., R. D. Qber, L. B. Schaaf, and D. G. Scott. 1980. Comparison of second-day pickup with numbers estimated by Pollution Committee counting guidelines: Barkley Lake study. Proceedings of the Annual Conference Southeastern Association of Fish and Wildlife Agencies 33(1979):738–752.

Henley, J. P. 1967. Evaluation of rotenone sampling with scuba gear. Proceedings of the Annual Conference Southeastern Association of Game and Fish Commissioners 20(1966):439–446.

Hill, D. M. 1983. Fish kill investigation procedures. Pages 261–274 *in* L. A. Nielsen and D. L. Johnson, editors. Fisheries techniques. American Fisheries Society, Bethesda, Maryland.

Huang J., Y. Cao, and K. S. Cummings. 2011. Assessing sampling adequacy of mussel diversity surveys in wadeable Illinois streams. Journal of the North American Benthological Society 30:923–934.

Johnson, P. D., A. E. Bogan, K. M. Brown, N. M. Burkhead, J. R. Cordeiro, J. T. Garner, P. D. Hartfield, D. A. W. Lepitzki, G. L. Mackie, E. Pip, T. A. Tarpley, J. S. Tiemann, N. V. Whelan, and E. E. Strong. 2013. Conservation status of freshwater gastropods of Canada and the United States. Fisheries 38:247–282.

Krumholz, L. A. 1944. A check on the fin-clipping method for estimating fish populations. Papers of the Michigan Academy of Science, Arts and Letters 29:281–291.

Krumholz, L. A. 1948. The use of rotenone in fisheries research. Journal of Wildlife Management 12:305–317.

Krumholz, L. A. 1950. Some practical considerations in the use of rotenone in fisheries research. Journal of Wildlife Management 14:413–424.

Leonard, J. W. 1939. Notes on the use of derris as a fish poison. Transactions of the American Fisheries Society 68:269–280.

Meyer, F. P., and L. A. Barclay, editors. 1990. Field manual for the investigation of fish kills. U.S. Fish and Wildlife Service, Resource Publication 177, Washington, D.C.

Murphy, B. R., D. W. Willis, and T. A. Springer. 1991. The relative weight index in fisheries management: status and needs. Fisheries 16(2):30–38.

Neumann, R. M., C. S. Guy, and D. W. Willis. 2012. Length, weight, and associated indices. Pages 637–676 *in* A. V. Zale, D. L. Parrish, and T. M. Sutton, editors. Fisheries techniques, 3rd edition. American Fisheries Society, Bethesda, Maryland.

Neves, R. J., and S. N. Moyer. 1988. Evaluation of techniques for age determination of freshwater mussels (Unionidae). American Malacological Bulletin 6:179–188.

Newton T. J., and W. G. Cope. 2007. Biomarker responses of unionid mussels to environmental contaminants. Pages 257–284 *in* J. L. Farris and J. H. Van Hassel, editors. Freshwater bivalve ecotoxicology. SETAC Press, Pensacola, Florida and Taylor & Francis, Boca Raton, Florida.

NOAA (National Oceanic and Atmospheric Administration. 1999. Discounting and the treatment of uncertainty in natural resource damage assessment. NOAA, Damage Assessment Center, Technical Paper 99-1, Silver Spring, Maryland.

Parker, R. O., Jr. 1970. Surfacing of dead fish following application of rotenone. Transactions of the American Fisheries Society 99:805–807.

Ricker, W. E. 1975. Computation and interpretation of biological statistics of fish populations. Bulletin of the Fisheries Research Board of Canada 191.

Rohlf, F. J., and R. R. Sokal. 1969. Statistical tables. W. H. Freeman and Company, San Francisco, California.

Schneider, J. C., P. W. Laarman, and H. Gowing. 2000. Manual of fisheries survey methods II. Michigan Department of Natural Resources, Fisheries Division, Lansing.

Shireman, J. V., D. E. Colle, and D. F. DuRant. 1981. Efficiency of rotenone sampling with large and small block nets in vegetated and open-water habitats. Transactions of the American Fisheries Society 110:77–80.

Southwick, R. I., and A. J. Loftus, editors. 2003. Investigation and monetary values of fish and freshwater mussel kills. American Fisheries Society, Special Publication 30, Bethesda, Maryland.

Smith, D. R. 2006. Survey design for detecting rare freshwater mussels. Journal of the North American Benthological Society 25:701–711.

Strayer, D. L., and D. R. Smith. 2003. A guide to sampling freshwater mussel populations. American Fisheries Society, Monograph 8, Bethesda, Maryland.

Thiel, P. A., and A. W. Fritz. 1993. Mussel harvest and regulations in the upper Mississippi River system. Pages 11–18 *in* K. S. Cummings, A. C. Buchanan, and L. M. Koch, editors. Conservation and management of freshwater mussels. Upper Mississippi River Conservation Committee, Rock Island, Illinois.

Tiemann, J. S., K. S. Cummings, and C. A. Mayer. 2009. Timed search technique used to evaluate freshwater mussel (Bivalvia: Unionidae) species richness in headwater streams: is a single one-hour visit enough? Journal of Freshwater Ecology 24:85–92.

Turgeon, D. D., A. E. Bogan, E. V. Coan, F. G. Hochberg, W. G. Lyons, P. M. Mikkelsen, J. F. Quinn, Jr., C. F. E. Roper, G. Rosenberg, B. Roth, A. Scheltema, M. J. Sweeney, F. G. Thompson, M. Vecchione, and J. D. Williams. 1998. Common and scientific names of aquatic invertebrates from the United States and Canada: mollusks, 2nd edition. American Fisheries Society, Special Publication 26, Bethesda, Maryland.

USEPA (U.S. Environmental Protection Agency). 2000a. Stressor identification guidance document. USEPA, 822/B-00/025, Washington, D.C.

USEPA (U.S. Environmental Protection Agency). 2000b. Methods for measuring the toxicity and bioaccumulation of sediment-associated contaminants with freshwater invertebrates, 2nd edition. USEPA, 600/R-99/064, Washington, D.C.

USOFR (U.S. Office of the Federal Register). 1990. Natural resource damage assessments.

Code of Federal Regulations, Title 15, Part 990. U.S. Government Printing Office, Washington, D.C.

USOFR (U.S. Office of the Federal Register). 1996. Natural resource damage assessments. Code of Federal Regulations, Title 43, Part 11. U.S. Government Printing Office, Washington, D.C.

Vaughn, C., and C. C. Hakenkamp. 2001. The functional role of burrowing bivalves in freshwater ecosystems. Freshwater Biology 46:1431–1446.

Vaughn, C. C. 2010. Biodiversity losses and ecosystem function in freshwaters: emerging conclusions and research directions. Bioscience 60:25–35.

Vaughn, C. C. 2017. Ecosystem services provided by freshwater mussels. Hydrobiologia, DOI 10.1007/s10750-017-3139-x.

Williams, J. D., A. E. Bogan, R. S. Butler, K. S. Cummings, J. T. Garner, J. L. Harris, N. A. Johnson, and G. T. Watters. 2017. A revised list of the freshwater mussels (Mollusca: Bivalvia: Unionida) of the United States and Canada. Freshwater Mollusks Biology and Conservation 20(2):32–58.

Williams, J. D., M. L. Warren, Jr., K. S. Cummings, J. L. Harris, and R. J. Neves. 1993. Conservation status of freshwater mussels of the United States and Canada. Fisheries 18(9):6–22.

Appendix A: Hatchery Production Costs of Fish

Note: In place of an errata sheet listing corrections, this is revision 1.1 of the original Appendix A of the 2017 version of this publication. It provides adjusted values for larger fish of some species using interpolation (consistent with the methodology in Southwick and Loftus 2003), which had been inadvertently excluded from the first version. Additionally, values for Common Carp were adjusted by removing hatchery costs for cultured decorative koi from the calculations (consistent with Southwick and Loftus 2003).

The data that follow are average costs of hatchery-raised fish organized by region, species, and size. They are derived from a 2016 survey of public, private, and tribal hatcheries, based on 328 responses received (survey details are in Appendix C). The replacement costs in this publication reflect hatchery costs in the contiguous United States.

Families of fish are listed in phylogenetic order according to the convention used by the American Fisheries Society). Within groups, species are listed alphabetically by common name. Data are expressed as cost per fish unless indicated otherwise. The costs given should not be used for rare, endangered, or threatened species. An asterisk (*) preceding a fish name indicates that one (or more) species—either in its entirety or as a subspecies or population—may be included, or is under review for inclusion, in the list of endangered and threatened wildlife under the U.S. Endangered Species Act of 1973, as amended. Appendix B has a more detailed listing of these protected species.

Costs are listed by inch, where each size increment given includes fish of that size down to the next increment below. For example, the cost given for a 7–8-in Hickory Shad *Alosa mediocris* would include all fish from 7.01 in up to 8.0 in. In cases where values reflect inflation-adjusted prices from previous editions of this publication, the categories used in the 2003 version (Southwick and Loftus 2003) are retained.

Data are stratified by geographic region (Figure A.1) to accommodate climatic and other regional variations that affect the cost of fish culture. When calculating replacement values for fish, use the most localized data available. If local hatchery data are available, use these data; if not, use regional data, and finally, average cost estimates for all regions combined (listed in the "All" column). In areas not well represented by survey hatchery data, such as Canada, use values from the most applicable region (e.g., Northeast region for the Maritime Provinces, etc.).

Some species were not represented in the 2016 survey of hatcheries. In these instances, the inflation-adjusted values (using the producer price index) published in the 2003 edition are listed and noted in italics.

The regions correspond to administrative regions of the U.S. Fish and Wildlife Service (FWS; Alaska is included with the Pacific region for analysis) plus Canadian provinces in close geographic proximity, as follows (Figure A.1): Pacific (P), FWS Regions 1 and 7; Mountain–Prairie (MP), FWS Region 6; Southwest (SW), FWS Region 2; Great Lakes–Big Rivers (GL), FWS Region 3; Southeast (SE), FWS Region 4; and Northeast (NE), FWS Region 5.

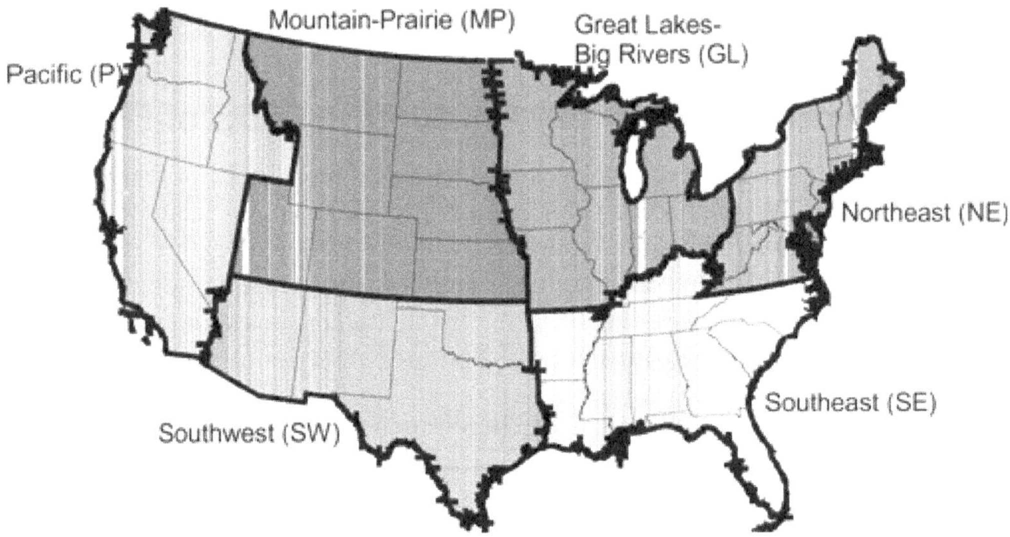

FIGURE A.1. Regions of the United States used to stratify hatchery replacement costs of fish. States and Canadian provinces in each region are as follows.

P (Pacific): Alaska, British Columbia, California, Hawaii, Idaho, Nevada, Oregon, Washington.

MP (Mountain-Prairie): Alberta, Colorado, Kansas, Manitoba, Montana, Nebraska, North Dakota, Saskatchewan, South Dakota, Utah, Wyoming.

SW (Southwest): Arizona, New Mexico, Oklahoma, Texas.

GL (Great Lakes-Big Rivers): Illinois, Indiana, Iowa, Michigan, Minnesota, Missouri, Ohio, Ontario, Wisconsin.

SE (Southeast): Alabama, Arkansas, Florida, Georgia, Kentucky, Louisiana, Mississippi, North Carolina, South Carolina, Tennessee.

NE (Northeast): Connecticut, Delaware, District of Columbia, Maine, Maryland, Massachusetts, New Brunswick, New Hampshire, New Jersey, New York, Newfoundland, Nova Scotia, Quebec, Pennsylvania, Rhode Island, Vermont, Virginia, West Virginia.

Size (inches)	P	MP	SW	GL	SE	NE	All
			ACIPENSERIFORMES				
			Acipenseridae—sturgeons				

*Atlantic Sturgeon . . . . . *Acipenser oxyrinchus*
Green Sturgeon . . . . . *Acipenser medirostris*
*Lake Sturgeon . . . . . *Acipenser fulvescens*
*Pallid Sturgeon . . . . . *Scaphirhynchus albus*
*Shortnose Sturgeon . . . . . *Acipenser brevirostrum*
*Shovelnose Sturgeon . . . . . *Scaphirhynchus platorynchus*
*White Sturgeon . . . . . *Acipenser transmontanus*

Size (inches)	P	MP	SW	GL	SE	NE	All
0–1	...	...	...	...	...	...	*$0.04*
1–2	...	...	...	...	...	...	*$0.93*
2–3	...	...	...	...	...	...	*$1.82*
3–4	...	...	...	...	...	...	*$2.69*
4–5	...	...	...	$2.15	...	...	*$3.58*
5–6	...	...	...	$2.15	...	...	*$4.06*
6–7	...	...	...	...	...	...	*$13.76*
7–8	...	...	...	$5.01	...	...	*$33.18*
8–9	...	...	...	...	...	...	*$33.76*
9–10	...	...	...	...	...	...	*$34.34*
10–11	...	...	...	...	...	...	*$34.65*
11–12	...	...	...	...	...	...	*$46.45*
12–13	...	...	...	...	...	...	*$66.35*
per pound >13							*$154.20*

Polyodontidae—paddlefishes

*Paddlefish . . . . . *Polyodon spathula*

Size (inches)	P	MP	SW	GL	SE	NE	All
0–1	...	...	$0.50	...	...	...	$0.50
1–2	...	...	$1.00	...	...	...	$1.00
2–3	...	...	$2.00	...	...	...	$2.00
3–4	...	...	$3.00	...	...	...	$3.00
4–5	...	...	$4.00	...	...	...	$4.00
5–6	...	...	$5.00	...	...	...	$5.00
6–7	...	...	$6.00	...	...	...	$6.00
7–8	...	...	$7.00	...	...	...	$7.00

Italicized value = inflation-adjusted 2003 values.
* = species or subspecies is federally listed as T&E in a portion of its range.
Region abbreviations are described in caption of Figure A.1.

Size (inches)	Region P	MP	SW	GL	SE	NE	All
8–9	...	...	$8.00	...	...	...	$8.00
9–10	...	...	$9.00	$12.57	...	...	$10.79
10–11	...	...	$10.00	$40.00	...	...	$25.00
11–12	...	...	$11.00	$40.00	...	...	$25.50
12–13	...	...	$12.00	$40.00	...	...	$26.00
13–14	...	...	...	$40.00	...	...	$40.00
per pound >14							$21.53

LEPISOSTEIFORMES

Lepisosteidae—gars

Alligator Gar . . . . *Atractosteus spatula*
Florida Gar. . . . . *Lepisosteus platyrhincus*
Longnose Gar. . . . . *Lepisosteus osseus*
Shortnose Gar . . . . *Lepisosteus platostomus*
Spotted Gar . . . . *Lepisosteus oculatus*

Size (inches)	P	MP	SW	GL	SE	NE	All
0–1	...	...	$0.50	...	...	...	$0.50
1–2	...	...	$1.00	...	...	...	$1.00
2–3	...	...	$2.00	...	...	...	$2.00
3–4	...	...	$3.00	...	...	...	$3.00
4–5	...	...	$4.00	...	...	...	$4.00
5–6	...	...	$5.00	...	...	...	$5.00
6–7	...	...	$6.00	...	...	...	$6.00
7–8	...	...	$7.00	...	...	...	$7.00
8–9	...	...	$8.00	...	...	...	$8.00
9–10	...	...	$9.00	...	...	...	$9.00
10–11	...	...	$10.00	...	...	...	$10.00
11–12	...	...	$11.00	...	...	...	$11.00
12–13	...	...	$12.00	...	...	...	$12.00
per pound >13							$58.33

AMIIFORMES

Amiidae—bowfins

Bowfin. . . . . *Amia calva*

	All
per pound	*$0.45*

Size (inches)	Region P	MP	SW	GL	SE	NE	All

HIODONTIFORMES

Hiodontidae—mooneyes

Goldeye. . . . . *Hiodon alosoides*
Mooneye. . . . . *Hiodon tergisus*

Size (inches)	P	MP	SW	GL	SE	NE	All
0–3	...	...	...	...	...	...	*$0.17*
3–6	...	...	...	...	...	...	*$0.37*
6–8	...	...	...	...	...	...	*$0.73*
8–10	...	...	...	...	...	...	*$1.09*
10–12	...	...	...	...	...	...	*$1.21*
per pound >12							*$2.14*

ANGUILLIFORMES

Anguillidae—freshwater eels

American Eel . . . . *Anguilla rostrata*

Size (inches)	P	MP	SW	GL	SE	NE	All
per fish							*$3.14*

CLUPEIFORMES

Clupeidae—herrings

Alewife . . . . *Alosa pseudoharengus*
Blueback Herring . . . . *Alosa aestivalis*

Size (inches)	P	MP	SW	GL	SE	NE	All
0–3	...	...	...	...	...	...	*$0.23*
3–6	...	...	...	...	...	...	*$0.50*
6–7	...	...	...	...	...	...	*$0.69*
7–8	...	...	...	...	...	...	*$0.88*
8–9	...	...	...	...	...	...	*$1.07*
per pound >9							*$3.86*

Alabama Shad . . . . *Alosa alabamae*
American Shad . . . . *Alosa sapidissima*
Hickory Shad . . . . *Alosa mediocris*

Size (inches)	P	MP	SW	GL	SE	NE	All
0–3	...	...	...	...	...	$0.10	*$0.23*
3–6	...	...	...	...	...	...	*$0.50*

Size (inches)	Region P	MP	SW	GL	SE	NE	All
6–7	...	...	...	...	...	...	*$0.76*
7–8	...	...	...	...	...	...	*$0.88*
8–9	...	...	...	...	...	...	*$1.13*
9–10	...	...	...	...	...	...	*$1.25*
10–11	...	...	...	...	...	...	*$1.59*
11–12	...	...	...	...	...	...	*$1.67*
12–13	...	...	...	...	...	...	*$1.88*
per pound >13							*$1.88*

Atlantic Menhaden . . . . *Brevoortia tyrannus*
Finescale Menhaden . . . . *Brevoortia gunteri*
Gizzard Shad . . . . *Dorosoma cepedianum*
Gulf Menhaden . . . . *Brevoortia patronus*
Skipjack Herring . . . . *Alosa chrysochloris*
Threadfin Shad . . . . *Dorosoma petenense*
Yellowfin Menhaden . . . . *Brevoortia smithi*

Size (inches)	P	MP	SW	GL	SE	NE	All
2–3	...	...	...	...	$0.35	...	$0.35
3–4	...	...	...	...	$0.70	...	$0.70
4–5	...	...	...	...	$1.05	...	$1.05
5–6	...	...	...	...	$1.40	...	$1.40
6–7	...	...	...	...	$1.75	...	$1.75
7–8	...	...	...	...	$1.75	...	$1.75
8–9	...	...	...	...	$1.75	...	$1.75
9–10	...	...	...	...	$1.75	...	$1.75
per pound >10							$4.50

Engraulidae—anchovies

Bay Anchovy . . . . *Anchoa mitchilli*

Size (inches)	P	MP	SW	GL	SE	NE	All
per fish							*$0.12*

CYPRINIFORMES

Cyprinidae—carps and minnows

Common Carp . . . . *Cyprinus carpio*
*stonerollers . . . . *Campostoma* spp.

Size (inches)	P	MP	SW	GL	SE	NE	All
0–6	...	...	...	...	...	...	*$0.12*
6–8	...	...	...	...	...	...	*$0.24*

Size (inches)	Region P	MP	SW	GL	SE	NE	All
6–8	...	...	...	...	...	...	*$0.24*
8–9	...	...	...	...	...	...	*$0.28*
9–10	...	...	...	...	...	...	*$0.36*
10–11	...	...	...	...	...	...	*$0.42*
per pound > 11							*$0.52*
Fathead Minnow . . . *Pimephales promelas*							
0–2	...	...	$0.04	$0.03	$0.07	...	$0.04
2–3	...	...	...	...	$0.07	...	$0.07
per pound >3							$5.57
Golden Shiner . . . *Notemigonus crysoleucas*							
0–6	...	...	...	...	...	...	*$0.28*
6–7	...	...	...	...	...	...	*$0.81*
7–12	...	...	...	...	...	...	*$1.33*
per pound >12							*$2.36*
Grass Carp, diploid . . . *Ctenopharyngodon idella*							
0–1	...	$0.01	...	...	...	...	$0.01
1–2	...	...	...	...	...	...	$0.37
2–3	...	...	...	...	...	...	$0.73
3–4	...	$0.18	...	...	$2.00	...	$1.09
4–5	...	$4.12	...	...	$2.00	...	$3.06
5–6	...	$4.14	...	...	...	...	$4.14
6–7	...	$8.00	...	...	...	...	$8.00
7–8	...	$9.00	...	...	...	...	$9.00
8–9	...	$10.67	$8.00	$8.84	$8.29	...	$9.42
9–10	...	$10.67	$8.00	$8.84	$8.29	...	$9.83
10–11	...	$10.83	$12.08	$9.18	$8.91	...	$10.25
11–12	...	$11.00	$16.17	$9.51	$11.00	$17.27	$12.99
12–13	...	...	$20.25	$9.84	...	$17.27	$15.79
13–14	...	...	$20.25	$10.17	...	$17.27	$15.90
14–15	...	...	...	$10.50	...	$17.27	$16.48
15–16	...	...	...	$10.83	...	...	$17.07
16–17	...	...	...	$12.67	...	...	$17.66
17–18	...	...	...	$14.50	...	...	$18.24
18–19	...	...	...	$16.33	...	...	$18.83

Size (inches)	Region P	MP	SW	GL	SE	NE	All
19–20	...	...	...	$18.17	...	...	$19.41
20–21	...	...	...	$20.00	...	...	$20.00
per pound >21							$3.80
	Grass Carp, triploid. . . . . . *Ctenopharyngodon idella*						
3–4	...	...	...	...	$10.00	...	$10.00
4–5	...	...	...	...	$10.00	...	$10.00
5–6	...	...	...	...	$10.00	$6.50	$10.08
6–7	...	...	...	...	$10.50	$8.00	$10.16
7–8	...	...	...	...	$11.00	$8.00	$10.24
8–9	...	$10.00	$12.50	$10.25	$11.36	$9.00	$10.62
9–10	...	$10.00	$12.50	$10.44	$11.72	$11.50	$11.23
10–11	...	$10.00	$13.00	$11.38	$12.08	$15.34	$12.36
11–12	...	$11.00	$13.50	$12.85	$12.44	$16.89	$13.34
12–13	...	$15.00	$13.50	$14.71	$14.00	$17.77	$15.00
13–14	...	$15.00	$13.50	$14.71	$15.00	$19.22	$15.49
14–15	...	...	$14.50	$14.71	...	$20.67	$16.63
per pound >14							$8.27
	Other cyprinids						
per fish							*$0.12*
	Catostomidae—suckers						
	Bigmouth Buffalo . . . . . . *Ictiobus cyprinellus* Black Buffalo . . . . . . *Ictiobus niger* Smallmouth Buffalo . . . . . . *Ictiobus bubalus*						
0–4	...	...	...	...	...	...	*$0.20*
4–6	...	...	...	...	...	...	*$0.36*
6–9	...	...	...	...	...	...	*$0.48*
9–10	...	...	...	...	...	...	*$0.65*
10–11	...	...	...	...	...	...	*$0.76*
11–12	...	...	...	...	...	...	*$0.93*
per pound >12							*$0.90*

Alabama Hog Sucker . . . . . . *Hypentelium etowanum*
Black Redhorse . . . . . . *Moxostoma duquesnei*
Blacktail Redhorse. . . . . . *Moxostoma poecilurum*

Blue Sucker . . . . . . *Cycleptus elongatus*
Eastern Creek Chubsucker . . . . . . *Erimyzon oblongus*
Golden Redhorse . . . . . . *Moxostoma erythrurum*
Lake Chubsucker . . . . . . *Erimyzon sucetta*
Longnose Sucker . . . . . . *Catostomus catostomus*
Mountain Sucker . . . . . . *Catostomus platyrhynchus*
Northern Hog Sucker . . . . . . *Hypentelium nigricans*
River Redhorse . . . . . . *Moxostoma carinatum*
Robust Redhorse . . . . . . *Moxostoma robustum*
Shorthead Redhorse . . . . . . *Moxostoma macrolepidotum*
Silver Redhorse . . . . . . *Moxostoma anisurum*
Spotted Sucker . . . . . . *Minytrema melanops*
White Sucker . . . . . . *Catostomus commersonii*

Size (inches)	P	MP	SW	GL	SE	NE	All
0–1	...	...	...	...	...	...	...
1–2	...	...	...	...	...	...	...
2–3	...	...	...	...	...	$0.70	$0.70
3–4	...	...	...	...	...	$0.85	$0.85
4–5	...	...	...	...	...	$1.00	$1.00
5–6	...	...	...	...	...	$2.00	$2.00
6–7	...	...	...	...	...	...	$2.88
7–8	...	...	...	...	$3.75	...	$3.75
per pound >8							$15.88

Quillback . . . . . . *Carpiodes cyprinus*
River Carpsucker . . . . . . *Carpiodes carpio*

Size (inches)	P	MP	SW	GL	SE	NE	All
0–3	...	...	...	...	...	...	*$0.09*
4–6	...	...	...	...	...	...	*$0.11*
7–8	...	...	...	...	...	...	*$0.20*
9–10	...	...	...	...	...	...	*$0.27*
10–11	...	...	...	...	...	...	*$0.36*
11–12	...	...	...	...	...	...	*$0.42*
per pound >12							*$0.50*

CHARACIFORMES

Characidae—tetras

Mexican Tetra . . . . . . *Astyanax mexicanus*

Size (inches)	All
per fish	*$0.12*

Size (inches)	Region P	MP	SW	GL	SE	NE	All

SILURFORMES

Ictaluridae—North American catfishes

Blue Catfish . . . . *Ictalurus furcatus*
Channel Catfish . . . . *ctalurus punctatus*
Flathead Catfish . . . . *Pylodictis olivaris*
White Catfish . . . . *Ameiurus catus*

Size (inches)	P	MP	SW	GL	SE	NE	All
0–1	...	...	$0.51	...	$0.05	...	$0.27
1–2	$0.50	$0.13	$0.54	$0.13	$0.10	$1.20	$0.55
2–3	$1.00	$0.45	$0.58	$0.20	$0.36	$1.24	$0.59
3–4	$1.50	$0.52	$0.62	$0.61	$0.44	$1.28	$0.74
4–5	$2.00	$0.68	$0.67	$0.64	$0.53	$1.64	$0.90
5–6	$2.50	$0.76	$0.71	$0.67	$0.61	$2.02	$1.05
6–7	$3.00	$1.03	$0.81	$1.00	$0.69	$2.14	$1.20
7–8	$3.50	$1.03	$0.94	$1.13	$0.74	$2.23	$1.35
8–9	...	$1.12	$1.35	$1.39	$1.24	$2.32	$1.51
9–10	...	$1.20	$1.75	$1.50	$1.28	$2.32	$1.92
10–11	...	$1.29	$1.79	$1.61	$1.33	$3.15	$2.05
11–12	...	$1.38	$2.38	$1.81	$1.38	$3.15	$2.18
12–13	...	...	$3.09	$2.52	$1.50	$3.77	$2.47
13–14	...	...	$3.79	$2.52	$1.50	$4.38	$2.76
14–15	...	...	$4.50	$4.29	...	$5.00	$4.52
15–16	...	...	$5.00	...	...	...	$5.00
16–17	...	...	$6.00	...	...	...	$6.00
17–18	...	...	$7.50	...	...	...	$7.50
18–19	...	...	$10.00	...	...	...	$10.00
19–20	...	...	$12.50	...	...	...	$12.50
20–21	...	...	$15.00	...	...	...	$15.00
21–22	...	...	$17.00	...	...	...	$17.00
22–23	...	...	$19.00	...	...	...	$19.00
23–24	...	...	$20.00	...	...	...	$20.00
24–25	...	...	$25.00	...	...	...	$25.00
25–26	...	...	$26.00	...	...	...	$26.00
per pound >26							$3.71

Black Bullhead . . . . *Ameiurus melas*
Brown Bullhead . . . . *Ameiurus nebulosus*
Flat Bullhead . . . . *Ameiurus platycephalus*
*madtoms . . . . *Noturus* spp.
Yellow Bullhead . . . . *Ameiurus natalis*

Size (inches)	Region P	MP	SW	GL	SE	NE	All
0–2	...	*$0.25*	...	...	...	...	*$0.33*
2–3	...	*$0.50*	...	...	...	*$0.50*	*$0.66*
3–4	...	*$0.50*	...	*$0.50*	...	*$0.68*	*$0.81*
4–5	...	*$0.75*	...	*$0.50*	...	*$0.90*	*$1.01*
5–6	...	*$1.00*	...	*$0.50*	...	*$0.90*	*$1.10*
6–7	...	...	...	...	...	*$1.63*	*$2.16*
7–8	...	...	...	...	...	*$1.80*	*$2.39*
8–9	...	...	...	...	...	*$2.30*	*$3.05*
9–10	...	...	...	...	...	*$3.00*	*$3.98*
per pound >10							*$6.29*

SALMONIFORMES

Salmonidae—trouts and salmons

*Atlantic Salmon .... *Salmo salar*
*Chinook Salmon .... *Oncorhynchus tshawytscha*
Chum Salmon .... *Oncorhynchus keta*
Coho Salmon............................... *Oncorhynchus kisutch*
Pink Salmon............................. *Oncorhynchus gorbuscha*
*Sockeye Salmon (=kokanee) .... *Oncorhynchus nerka*

Size (inches)	P	MP	SW	GL	SE	NE	All
0–1	...	...	...	...	...	$0.01	$0.01
1–2	$0.13	$0.09	...	...	...	$0.01	$0.08
2–3	$0.13	$0.20	...	...	...	$0.03	$0.12
3–4	$0.22	$0.28	...	$0.43	...	$0.44	$0.34
4–5	$0.39	$0.36	$0.51	$0.71	...	$0.49	$0.49
5–6	$0.47	$0.37	...	$0.99	...	$0.54	$0.59
6–7	$0.55	...	...	$1.70	...	$0.83	$1.02
7–8	$0.77	...	...	$1.70	...	$1.12	$1.20
8–9	...	...	...	$1.70	...	$1.55	$1.62
9–10	...	...	...	...	...	...	$2.72
10–11	...	...	...	...	...	...	$3.82
11–12	...	...	...	...	...	$4.92	$4.92
12–13	...	...	...	...	...	...	$7.75
13–14	...	...	...	...	...	...	$10.59
14–15	...	...	...	...	...	...	$13.43
15–16	...	...	...	...	...	...	$16.26
16–17	...	...	...	...	...	...	$19.10
17–18	...	...	...	...	...	...	$21.94
18–19	...	...	...	...	...	...	$24.77
19–20	...	...	...	...	...	...	$27.61

Size	Region						
(inches)	P	MP	SW	GL	SE	NE	All
20–21	...	...	...	...	...	...	$30.45
21–22	...	...	...	...	...	...	$33.29
22–23	...	...	...	...	...	...	$36.12
23–24	...	...	...	...	...	$38.96	$38.96
24–25	...	...	...	...	...	...	$41.80
25–26	...	...	...	...	...	$51.14	$44.63
per pound >26							$6.13

Apache Trout . . . . *Oncorhynchus apache*
Arctic Char. . . . . *Salvelinus alpinus*
Arctic Grayling . . . . *Thymallus arcticus*
Brook Trout . . . . *Salvelinus fontinalis*
Brown Trout. . . . . *Salmo trutta*
ciscoes and whitefishes. . . . . *Coregonus* spp.
*Cutthroat Trout . . . . *Oncorhynchus clarkii*
Gila Trout. . . . . *Oncorhynchus gilae*
Golden Trout . . . . *Oncorhynchus aguabonita*
Lake Trout . . . . *Salvelinus namaycush*
tiger trout . . . . *Salmo trutta* × *Salvelinus fontinalis*
whitefishes . . . . *Prosopium* spp.

Size (inches)	P	MP	SW	GL	SE	NE	All
0–1	$0.25	...	$0.05	...	$0.10	$0.19	$0.28
1–2	$0.35	$0.70	$0.08	$0.55	$0.25	$0.36	$0.37
2–3	$0.37	$0.95	$0.16	$0.57	$0.30	$0.65	$0.51
3–4	$0.39	$1.78	$0.25	$0.59	$0.37	$0.88	$0.83
4–5	$0.45	$1.97	$0.31	$2.11	$0.61	$1.09	$1.23
5–6	$0.52	$2.56	$0.39	$2.18	$0.78	$1.28	$1.41
6–7	$0.59	$2.67	...	$2.28	$0.96	$1.65	$1.72
7–8	$0.60	$3.14	...	$2.64	$1.14	$2.00	$2.11
8–9	$0.61	$3.16	$1.39	$2.96	$1.36	$2.46	$2.55
9–10	$0.63	$3.44	...	$3.27	$2.62	$2.79	$2.94
10–11	$0.64	$5.07	...	$3.42	$5.60	$3.59	$3.73
11–12	...	$5.63	...	$3.64	$5.60	$4.01	$4.16
12–13	...	$5.84	...	$3.85	$5.60	$4.80	$4.92
13–14	...	$6.06	...	$4.05	$7.17	$5.52	$5.61
14–15	...	...	...	...	...	$7.05	$7.05
15–16	...	...	...	...	...	$10.86	$10.86
16–17	...	...	...	...	...	$13.79	$13.79
17–18	...	...	...	...	...	$18.63	$18.63
18–19	...	...	...	...	...	$22.50	$22.50
19–20	...	...	...	...	...	$27.40	$27.40

Size	Region						
(inches)	P	MP	SW	GL	SE	NE	All
20–21	...	...	...	...	...	$32.64	$32.64
21–22	...	...	...	...	...	$39.01	$39.01
22–23	...	...	...	...	...	$45.60	$45.60
23–24	...	...	...	...	...	$54.36	$54.36
24–25	...	...	...	...	...	$60.03	$60.03
25–26	...	...	...	...	...	$60.52	$60.52
26–27	...	...	...	...	...	$61.00	$64.00
27–28	...	...	...	...	...	$66.00	$66.00
per pound >28							$7.64

Rainbow Trout. . . . . *Oncorhynchus mykiss*

Size (inches)	P	MP	SW	GL	SE	NE	All
0–1	$0.33	...	$0.08	$0.69	$0.10	$0.02	$0.27
1–2	$0.36	$0.31	$0.11	$0.96	$0.17	$0.40	$0.32
2–3	$0.39	$0.57	$0.17	$1.23	$0.25	$0.88	$0.49
3–4	$0.42	$1.05	$0.25	$1.49	$0.31	$0.88	$0.75
4–5	$0.53	$1.24	$0.32	$2.20	$0.55	$1.25	$1.07
5–6	$0.53	$1.61	$0.43	$2.38	$0.55	$1.38	$1.21
6–7	$0.72	$1.80	$0.47	$2.56	$0.56	$1.66	$1.49
7–8	$0.73	$2.21	$0.70	$2.61	$0.80	$1.89	$1.57
8–9	$0.89	$2.33	$0.81	$3.09	$0.96	$2.48	$2.11
9–10	$1.49	$2.46	$0.97	$3.57	$1.10	$2.80	$2.42
10–11	$1.50	$2.54	...	$3.81	$1.40	$3.55	$2.85
11–12	$1.51	$2.97	...	$4.06	$1.64	$3.91	$3.12
12–13	$2.29	$4.05	...	$4.50	$2.92	$5.02	$4.22
13–14	$3.18	$5.40	...	$4.95	$3.74	$5.84	$5.25
14–15	$3.40	...	...	$6.00	...	$7.54	$6.67
15–16	$4.40	...	...	...	...	$11.17	$10.27
16–17	...	...	...	...	...	$14.06	$14.06
17–18	...	...	...	...	...	$18.54	$18.54
18–19	...	...	...	...	...	$22.64	$22.64
19–20	...	...	...	...	...	$27.43	$27.43
20–21	$4.70	...	...	...	...	$33.07	$27.91
21–22	...	...	...	...	...	$39.37	$39.37
22–23	...	...	...	...	...	$45.06	$45.06
23–24	...	...	...	...	...	$54.73	$54.73
24–25	...	...	...	...	...	$59.00	$59.00
25–26	...	...	...	...	...	$61.09	$61.09
26–27	...	...	...	...	...	$63.42	$63.42
27–28	...	...	...	...	...	$66.00	$66.00
per pound >28							$7.02

Size (inches)	P	MP	SW	GL	SE	NE	All
				ESOCIIFORMES			
				Esocidae—pikes and mudminnows			

Chain Pickerel . . . . *Esox niger*
Grass Pickerel. . . . . *Esox americanus vermiculatus*
Northern Pike . . . . *Esox lucius*
Redfin Pickerel. . . . . *Esox americanus*

Size (inches)	P	MP	SW	GL	SE	NE	All
0–1	...	...	...	$0.01	...	...	$0.01
1–2	...	$0.07	$0.05	$0.27	...	...	$0.13
2–3	...	...	...	$0.66	...	...	$0.66
3–4	...	...	...	$1.04	...	...	$1.04
4–5	...	...	...	$3.00	...	...	$3.00
5–6	...	...	...	$3.00	...	...	$3.00
6–7	...	...	...	$5.33	...	...	$5.33
7–8	...	...	...	$7.67	...	...	$7.67
8–9	...	...	...	$10.00	$12.00	...	$10.00
9–10	...	...	...	$12.89	$12.00	...	$12.33
10–11	...	...	...	$15.78	$12.00	...	$14.67
11–12	...	...	...	$18.67	$12.00	...	$17.00
12–13	...	...	...	$20.78	...	...	$20.78
13–14	...	...	...	$22.89	...	...	$22.89
14–15	...	...	...	$25.00	...	...	$25.00
15–16	...	...	...	$25.00	...	...	$25.00
16–17	...	...	...	$25.00	...	...	$25.00
17–18	...	...	...	$25.00	...	...	$25.00
per pound >18							$19.28

Muskellunge. . . . . *Esox masquinongy*
tiger muskellunge . . . . *Esox lucius* × *Esox masquinongy*

Size (inches)	P	MP	SW	GL	SE	NE	All
0–1	...	...	...	...	...	...	$0.90
1–2	...	...	...	...	...	...	$2.03
2–3	...	...	...	...	...	...	$2.92
3–4	...	...	...	...	...	...	$3.85
4–5	...	...	...	...	...	...	$3.99
5–6	...	...	...	...	...	...	$6.45
6–7	...	...	...	$20.50	...	...	$8.91
7–8	...	...	...	$21.33	...	...	$11.37
8–9	...	...	...	$22.16	...	...	$15.00
9–10	...	...	...	$22.77	...	...	$17.00

Size (inches)	P	MP	SW	GL	SE	NE	All
10–11	...	...	...	$23.39	...	...	$19.00
11–12	...	...	...	$24.01	...	...	$21.00
12–13	...	...	...	$24.51	...	...	$23.00
13–14	...	...	...	$25.00	...	...	$25.00
per pound >14							$50.66

mudminnows. . . . . *Umbra* spp.

per fish	*$0.12*

ATHERINIFORMES*

Atherinopsidae—New World silversides

Brook Silverside. . . . . *Labidesthes sicculus*
Inland Silverside . . . . *Menidia beryllina*
*Waccamaw Silverside . . . . *Menidia extensa*

per fish	*$0.12*

ORDER CYPRINODONTIFORMES

Poeciliidae—livebearers

Eastern Mosquitofish . . . . *Gambusia holbrooki*
Western Mosquitofish. . . . . *Gambusia affinis*

	P	MP	SW	GL	SE	NE	All
per fish					$0.80		$0.80

PERCIFORMES

Moronidae—temperate basses

Striped Bass . . . . *Morone saxatilis*

Size (inches)	P	MP	SW	GL	SE	NE	All
0–1	...	$0.01	...	...	$0.01	...	$0.01
1–2	...	$1.00	$0.17	...	$0.22	$1.00	$0.39
2–3	...	$2.00	...	...	$1.22	$1.75	$1.53
3–4	...	$2.00	...	$1.37	$1.53	$2.25	$1.72
4–5	...	$2.85	$1.82	$1.72	$1.98	$3.25	$2.18
5–6	...	$3.35	$2.13	$2.70	$2.25	$3.75	$2.76
6–7	...	$4.13	$2.90	$3.39	$2.63	$5.75	$3.36

Size (inches)	P	MP	SW	GL	SE	NE	All
7–8	...	$4.63	$2.90	$4.01	$3.01	$7.00	$3.73
8–9	...	$5.88	$3.25	$5.62	$3.48	$8.00	$5.28
9–10	...	$6.38	$3.25	$5.84	$3.73	$9.00	$5.58
10–11	...	$10.00	$5.50	...	$3.73	$10.00	$6.59
11–12	...	...	$5.50	...	...	$11.00	$6.65
12–13	...	...	$11.00	...	...	$12.00	$11.50
13–14	...	...	$11.00	...	...	$13.00	$12.00
14–15	...	...	...	...	...	$14.00	$14.00
15–16	...	...	...	...	...	$15.00	$15.00
per pound >16							$8.19
Striped Bass hybrids.... *Morone saxatilis* × *Morone chrysops*							
0–1	...	...	...	$0.01	...	...	$0.01
1–2	...	...	$0.08	$0.06	$0.25	...	$0.15
2–3	...	$0.22	$0.20	$0.10	...	...	$0.18
3–4	...	$1.05	...	...	...	...	$1.05
4–5	...	$1.85	...	$0.64	...	...	$1.25
5–6	...	$2.15	...	...	...	...	$2.15
6–7	...	$2.45	...	...	...	...	$2.45
7–8	...	$2.45	...	...	...	...	$2.45
per pound >8							$8.19
White Bass .... *Morone chrysops* Yellow Bass .... *Morone mississippiensis*							
0–1	...	...	...	...	...	...	*$0.24*
1–2	...	$0.27	$0.02	...	...	...	*$0.48*
2–3	...	...	...	...	...	...	*$0.70*
3–4	...	...	...	...	...	...	*$0.94*
4–5	...	...	...	...	...	...	*$1.18*
5–6	...	...	...	...	...	...	*$1.42*
6–7	...	...	...	...	...	...	*$1.65*
7–8	...	...	...	...	...	...	*$1.88*
8–9	...	...	...	...	...	...	*$2.12*
9–10	...	...	...	...	...	...	*$2.36*
10–11	...	...	...	...	...	...	*$2.60*
11–12	...	...	...	...	...	...	*$2.83*
per pound >12							*$3.32*

Size (inches)	Region P	MP	SW	GL	SE	NE	All
	White Perch . . . . *Morone americana*						
0–1	...	...	...	...	...	...	*$0.17*
1–2	...	...	...	...	...	...	*$0.28*
2–3	...	...	...	...	...	...	*$0.48*
3–4	...	...	...	...	...	...	*$0.60*
4–5	...	...	...	...	...	...	*$0.88*
5–6	...	...	...	...	...	...	*$1.04*
6–7	...	...	...	...	...	...	*$1.21*
7–8	...	...	...	...	...	...	*$1.41*
8–9	...	...	...	...	...	...	*$1.58*
9–10	...	...	...	...	...	...	*$1.75*
10–11	...	...	...	...	...	...	*$1.94*
11–12	...	...	...	...	...	...	*$2.07*
per pound >12							*$2.02*

Centrarchidae—sunfishes

Size (inches)	P	MP	SW	GL	SE	NE	All
	Guadalupe Bass . . . . *Micropterus treculii* Largemouth Bass. . . . . *Micropterus salmoides* Redeye Bass . . . . *Micropterus coosae* Spotted Bass. . . . . *Micropterus punctulatus*						
0–1	...	...	...	...	$0.01	...	$0.01
1–2	$1.00	$0.64	$1.08	$0.07	$0.91	$0.85	$0.90
2–3	$2.00	$0.82	$1.17	$1.04	$1.24	$1.15	$1.15
3–4	$2.50	$0.86	$1.73	$1.30	$1.29	$1.80	$1.43
4–5	$5.00	$1.67	$2.09	$2.16	$1.69	$2.68	$2.27
5–6	$5.50	$2.24	$2.40	$2.68	$1.88	$2.93	$2.68
6–7	$6.00	$2.81	$4.32	$3.52	$3.32	$4.36	$3.79
7–8	$7.00	$3.57	$4.32	$3.71	$3.54	$4.86	$4.00
8–9	$8.00	$4.33	$5.13	$5.63	$4.28	$6.25	$5.66
9–10	...	$4.33	$5.13	$6.59	$4.28	$6.53	$6.68
10–11	...	$7.50	$6.25	$7.55	$6.88	$9.04	$7.86
11–12	...	$7.50	$7.00	$7.55	$6.88	$9.58	$7.98
12–13	...	...	...	...	...	$9.83	$8.10
13–14	...	...	...	...	...	$10.75	$8.13
14–15	...	...	...	...	...	$14.00	$14.00
15–16	...	...	...	...	...	$15.00	$15.00
per pound >16							$6.64

Size (inches)	Region P	MP	SW	GL	SE	NE	All
	Smallmouth Bass. . . . . *Micropterus dolomieu*						
0–1	...	...	...	...	...	$0.75	$0.75
1–2	...	$0.96	...	...	$0.48	$1.00	$1.40
2–3	...	$2.48	$2.39	$3.62	$1.40	...	$2.05
3–4	...	$4.00	...	$4.35	$2.10	...	$3.40
4–5	...	$4.00	...	$5.07	$3.90	$4.50	$4.63
5–6	...	$4.00	...	$5.33	...	$6.56	$5.31
6–7	...	$8.00	...	$9.27	...	$6.56	$7.94
7–8	...	$8.00	...	$9.27	...	...	$8.64
per pound >8							$33.46
	Black Crappie. . . . . *Pomoxis nigromaculatus* White Crappie . . . . *Pomoxis annularis*						
0–1	...	...	...	...	...	...	...
1–2	$0.75	$0.60	$0.70	$0.86	$0.87	$1.62	$0.83
2–3	$1.50	$0.73	$0.78	$0.98	$0.94	$1.78	$0.93
3–4	$2.25	$0.87	$0.87	$1.28	$1.35	$2.21	$1.35
4–5	$3.00	$1.49	$1.12	$1.68	$1.72	$2.64	$1.80
5–6	$3.75	$1.49	$2.25	$2.30	$2.47	$3.76	$2.54
6–7	...	$2.67	$2.38	$3.73	$2.50	$4.05	$3.44
7–8	...	$2.67	...	$4.11	...	$4.17	$3.73
8–9	...	$6.00	...	$4.50	...	$4.83	$5.00
9–10	...	$6.00	...	$4.50	...	$5.25	$5.25
10–11	...	...	...	$4.50	...	$7.50	$6.00
11–12	...	...	...	$7.50	...	$7.50	$7.50
12–13	...	...	...	$7.50	...	...	$7.50
13–14	...	...	...	$7.50	...	...	$7.50
14–15	...	...	...	$7.50	...	...	$7.50
per pound >8							$3.62

Banded Sunfish . . . . *Enneacanthus obesus*
Bluegill . . . . *Lepomis macrochirus*
Bluespotted Sunfish. . . . . *Enneacanthus gloriosus*
Dollar Sunfish . . . . *Lepomis marginatus*
Flier. . . . . *Centrarchus macropterus*
Green Sunfish . . . . *Lepomis cyanellus*
Longear Sunfish. . . . . *Lepomis megalotis*
Orangespotted Sunfish. . . . . *Lepomis humilis*

Size (inches)	Region P	MP	SW	GL	SE	NE	All
	Pumpkinseed . . . . . *Lepomis gibbosus*						
	Redbreast Sunfish . . . . . *Lepomis auritus*						
	Redear Sunfish . . . . . *Lepomis microlophus*						
	Rock Bass . . . . . *Ambloplites rupestris*						
	Shadow Bass . . . . . *Ambloplites ariommus*						
	Spotted Sunfish . . . . . *Lepomis punctatus*						
	Warmouth . . . . . *Lepomis gulosus*						
0–1	...	$0.23	...	$0.02	$0.14	...	$0.11
1–2	$0.75	$0.45	$0.43	$0.44	$0.40	$0.90	$0.44
2–3	$1.50	$0.58	$0.50	$0.56	$0.46	$0.92	$0.53
3–4	$2.13	$0.87	$0.67	$0.89	$0.60	$1.54	$0.84
4–5	$4.50	$1.41	$0.88	$1.34	$0.87	$1.88	$1.26
5–6	$5.63	$1.73	$1.15	$1.94	$1.49	$2.42	$1.90
6–7	...	$1.79	$2.16	$3.06	$1.92	$3.32	$2.68
7–8	...	...	$2.75	$3.16	$2.21	$3.75	$2.96
8–9	...	...	...	$4.08	$3.14	$4.25	$3.54
9–10	...	...	...	$5.00	$3.14	$5.00	$4.07
10–11	...	...	...	$5.00	$5.75	...	$5.38
11–12	...	...	...	$5.00	$5.75	...	$5.38
per pound >12							$3.71

Percidae—perches

Size (inches)	P	MP	SW	GL	SE	NE	All
	*darters . . . . . *Etheostoma* spp., *Percina* spp.						
	Yellow Perch . . . . . *Perca flavescens*						
0–1	...	...	...	$0.01	...	...	$0.45
1–2	...	$0.43	$0.31	$0.03	$1.08	...	$0.59
2–3	...	$0.48	$0.54	$0.77	$1.15	...	$0.81
3–4	...	$0.77	...	$1.24	$1.75	...	$1.36
4–5	...	$1.35	...	$1.67	$2.14	...	$1.82
5–6	...	$1.35	...	$1.96	$2.77	...	$2.16
6–7	...	$2.00	...	$2.85	$3.91	...	$3.10
7–8	...	$2.00	...	$3.57	$4.79	...	$3.83
8–9	...	...	...	$5.58	$5.83	...	$5.66
9–10	...	...	...	$5.58	$6.00	...	$5.71
10–11	...	...	...	$5.63	$6.00	...	$5.75
11–12	...	...	...	$5.63	$7.75	...	$6.33
12–13	...	...	...	...	$7.75	...	$7.75
per pound >13							$6.27

Size (inches)	Region						
	P	MP	SW	GL	SE	NE	All
	Sauger . . . . *Sander canadensis*						
	Walleye . . . . *Sander vitreus*						
0–1	...	$0.01	...	$0.01	...	...	$0.01
1–2	...	$0.17	$0.02	$0.04	...	$0.23	$0.12
2–3	...	$0.29	...	$0.24	...	$0.85	$0.34
3–4	...	$0.89	...	$1.68	...	$1.50	$1.56
4–5	...	$1.43	...	$1.78	...	$2.66	$2.02
5–6	...	$1.83	...	$2.84	...	$2.74	$2.71
6–7	...	$2.99	...	$2.86	...	$3.56	$3.03
7–8	...	$4.49	...	$3.76	...	$4.00	$3.57
8–9	...	$6.00	...	$4.23	...	$4.80	$4.58
9–10	...	$7.00	...	$5.87	...	...	$6.41
10–11	...	$9.00	...	$7.50	...	...	$8.25
11–12	...	$9.00	...	$7.50	...	...	$8.25
12–13	...	$15.00	...	...	...	...	$15.00
13–14	...	$15.00	...	...	...	...	$15.00
14–15	...	$15.00	...	...	...	...	$15.00
per pound >15							$12.32
	Sciaenidae—drums and croakers						
	Freshwater Drum . . . . *Aplodinotus grunniens*						
0–3	...	...	...	...	...	...	*$0.17*
3–6	...	...	...	...	...	...	*$0.28*
6–8	...	...	...	...	...	...	*$0.42*
8–10	...	...	...	...	...	...	*$0.60*
10–11	...	...	...	...	...	...	*$0.70*
11–12	...	...	...	...	...	...	*$0.82*
per pound >12							*$1.09*
	Red Drum (Redfish) . . . . *Sciaenops ocellatus*						
0–2	...	...	...	...	...	...	*$0.12*
2–3	...	...	...	...	...	...	*$0.24*
3–4	...	...	...	...	...	...	*$0.37*
4–5	...	...	...	...	...	...	*$0.84*
5–6	...	...	...	...	...	...	*$1.30*
6–7	...	...	$8.00	...	...	...	*$1.66*
per pound >7							*$9.75*

Size (inches)	Region P	MP	SW	GL	SE	NE	All
	Spotted Seatrout . . . . . . *Cynoscion nebulosus*						
per pound							*$29.31*
	Cichlidae—cichlids and tilapias						
	Blackchin Tilapia . . . . . . *Sarotherodon melanotheron* Blue Tilapia . . . . . . *Oreochromis aureus* Mozambique Tilapia . . . . . . *Oreochromis mossambicus* Redbelly Tilapia . . . . . . *Tilapia zillii* Spotted Tilapia . . . . . . *Tilapia mariae* tilapias . . . . . . unspecified and hybrids						
0–1	$0.55	...	$1.33	...	...	...	$0.94
1–2	$1.04	...	$1.33	$1.07	$0.83	...	$1.07
2–3	$2.00	...	$1.50	$1.41	$0.89	...	$1.33
3–4	$2.00	...	$1.50	$2.06	$0.95	...	$1.66
4–5	...	...	$2.50	$2.68	$1.20	...	$2.22
5–6	...	...	$2.50	...	$1.50	...	$3.56
6–7	...	...	...	...	$4.90	...	$4.90
7–8	...	...	...	...	$5.05	...	$5.05
8–9	...	...	...	...	$8.00	...	$8.00
per pound >7							$16.32

Appendix B: Threatened and Endangered Species of Fish

Summarized below are fish taxa that are included in the list of endangered and threatened wildlife under the U.S. Endangered Species Act of 1973. The summary was compiled from the U.S. Fish and Wildlife Service, Division of Endangered Species Web site on February 24, 2017. **Please check this source for any changes or updates to the list.** State and provincial governments also have statutes protecting rare, threatened, and endangered species that may be in addition to those listed here.

Species are listed alphabetically by the scientific name under which they were formally designated for protection.

Listing status codes:

E = endangered
T = threatened
P = proposed for listing
EXPE, XE = experimental population, essential EXPN; EXPE, XN = experimental population, nonessential
S = similarity of appearance to a threatened taxon

Note that some species may only be listed for a portion of its range or a segment of its population. Check official federal listing authorities for specific details.

Species	Common name	Status
Acanthorutilus handlirschi	Cicek (minnow)	E
Acipenser brevirostrum	Shortnose Sturgeon	E
Acipenser medirostris	Green Sturgeon	T
Acipenser mikadoi	Sakhalin Sturgeon	E
Acipenser naccarii	Adriatic Sturgeon	E
Acipenser oxyrinchus (=oxyrhynchus) desotoi	Atlantic Sturgeon (Gulf subspecies)	T
Acipenser oxyrinchus oxyrinchus	Atlantic Sturgeon	E, T
Acipenser sinensis	Chinese Sturgeon	E
Acipenser sturio	European Sturgeon	E
Acipenser transmontanus	White Sturgeon	E
Amblyopsis rosae	Ozark Cavefish	T
Anoxypristis cuspidata	Narrow Sawfish	E
Catostomus discobolus yarrowi	Zuni Bluehead Sucker	E
Catostomus santaanae	Santa Ana Sucker	T
Catostomus warnerensis	Warner Sucker	T
Chasmistes brevirostris	Shortnose Sucker	E
Chasmistes cujus	Cui-ui	E
Chasmistes liorus	June Sucker	E

Species	Common name	Status
Chrosomus saylori	Laurel Dace	E
Coreobagrus ichikawai	Nekogigi (catfish)	E
Cottus paulus (=pygmaeus)	Pygmy Sculpin	T
Cottus specus	Grotto Sculpin	E
Crenichthys baileyi baileyi	White River Springfish	E
Crenichthys baileyi grandis	Hiko White River Springfish	E
Crenichthys nevadae	Railroad Valley Springfish	T
Crystallaria cincotta	Diamond Darter	E
Cynoscion macdonaldi	Totoaba (seatrout or weakfish)	E
Cyprinella caerulea	Blue Shiner	T
Cyprinella formosa	Beautiful Shiner	T
Cyprinodon bovinus	Leon Springs Pupfish	E
Cyprinodon diabolis	Devils Hole Pupfish	E
Cyprinodon elegans	Comanche Springs Pupfish	E
Cyprinodon macularius	Desert Pupfish	E
Cyprinodon nevadensis mionectes	Ash Meadows Amargosa Pupfish	E
Cyprinodon nevadensis pectoralis	Warm Springs Pupfish	E
Cyprinodon radiosus	Owens Pupfish	E
Deltistes luxatus	Lost River Sucker	E
Dionda diaboli	Devils River Minnow	T
Elassoma alabamae	Spring Pygmy Sunfish	T
Empetrichthys latos	Pahrump Poolfish	E
Eremichthys acros	Desert Dace	T
Erimonax monachus	Spotfin Chub	XN, T
Erimystax cahni	Slender Chub	XN, T
Etheostoma boschungi	Slackwater Darter	T
Etheostoma chermocki	Vermilion Darter	E
Etheostoma chienense	Relict Darter	E
Etheostoma etowahae	Etowah Darter	E
Etheostoma fonticola	Fountain Darter	E
Etheostoma moorei	Yellowcheek Darter	E
Etheostoma nianguae	Niangua Darter	T
Etheostoma nuchale	Watercress Darter	E
Etheostoma okaloosae	Okaloosa Darter	T
Etheostoma percnurum	Duskytail Darter	E, XN
Etheostoma phytophilum	Rush Darter	E
Etheostoma rubrum	Bayou Darter	T
Etheostoma scotti	Cherokee Darter	T
Etheostoma sellare	Maryland Darter	E
Etheostoma sp.	Bluemask (=Jewel) Darter	E
Etheostoma spilotum	Kentucky Arrow Darter	T
Etheostoma susanae	Cumberland Darter	E
Etheostoma wapiti	Boulder Darter	E, XN

Species	Common name	Status
Eucyclogobius newberryi	Tidewater Goby	E
Gambusia gaigei	Big Bend Gambusia	E
Gambusia georgei	San Marcos Gambusia	E
Gambusia heterochir	Clear Creek Gambusia	E
Gambusia nobilis	Pecos Gambusia	E
Gasterosteus aculeatus williamsoni	Unarmored Threespine Stickleback	E
Gila bicolor ssp.	Hutton Tui Chub	T
Gila bicolor ssp. *mohavensis*	Mohave Tui Chub	E
Gila bicolor ssp. *snyderi*	Owens Tui Chub	E
Gila boraxobius	Borax Lake Chub	E
Gila cypha	Humpback Chub	E
Gila ditaenia	Sonora Chub	T
Gila elegans	Bonytail Chub	E
Gila intermedia	Gila Chub	E
Gila nigra	Headwater Chub	P
Gila nigrescens	Chihuahua Chub	T
Gila purpurea	Yaqui Chub	E
Gila robusta	Roundtail Chub	P
Gila robusta jordani	Pahranagat Roundtail Chub	E
Gila seminuda (=*robusta*)	Virgin River Chub	E
Huso dauricus	Kaluga Sturgeon	E
Huso huso	Beluga Sturgeon	T
Hybognathus amarus	Rio Grande Silvery Minnow	E, XN
Hymenophysa curta	Ayumodoki (loach)	E
Hypomesus transpacificus	Delta Smelt	T
Ictalurus pricei	Yaqui Catfish	T
Lepidomeda albivallis	White River Spinedace	E
Lepidomeda mollispinis pratensis	Big Spring Spinedace	T
Lepidomeda vittata	Little Colorado Spinedace	T
Meda fulgida	Spikedace	E
Menidia extensa	Waccamaw Silverside	T
Moapa coriacea	Moapa Dace	E
Notropis albizonatus	Palezone Shiner	E
Notropis buccula	Smalleye Shiner	E
Notropis cahabae	Cahaba Shiner	E
Notropis girardi	Arkansas River Shiner	T
Notropis mekistocholas	Cape Fear Shiner	E
Notropis oxyrhynchus	Sharpnose Shiner	E
Notropis simus pecosensis	Pecos Bluntnose Shiner	T
Notropis topeka (=*tristis*)	Topeka Shiner	E, XN
Noturus baileyi	Smoky Madtom	E, XN
Noturus crypticus	Chucky Madtom	E
Noturus flavipinnis	Yellowfin Madtom	XN, T

Species	Common name	Status
Noturus placidus	Neosho Madtom	T
Noturus stanauli	Pygmy Madtom	E, XN
Noturus trautmani	Scioto Madtom	E
Oncorhynchus (=Salmo) kisutch	Coho Salmon	E, T
Oncorhynchus (=Salmo) mykiss	Steelhead	E, XN, T
Oncorhynchus (=Salmo) nerka	Sockeye Salmon	E, T
Oncorhynchus (=Salmo) tshawytscha	Chinook Salmon	E, T
Oncorhynchus aguabonita whitei	Little Kern Golden Trout	T
Oncorhynchus apache	Apache Trout	T
Oncorhynchus clarki stomias	Greenback Cutthroat Trout	T
Oncorhynchus clarkii henshawi	Lahontan Cutthroat Trout	T
Oncorhynchus clarkii seleniris	Paiute Cutthroat Trout	T
Oncorhynchus gilae	Gila Trout	T
Oncorhynchus keta	Chum salmon	T
Pangasianodon gigas	Thailand Giant Catfish	E
Pangasius sanitwongsei	Catfish	E
Percina antesella	Amber Darter	E
Percina aurora	Pearl Darter	P
Percina aurolineata	Goldline Darter	T
Percina jenkinsi	Conasauga Logperch	E
Percina pantherina	Leopard Darter	T
Percina rex	Roanoke Logperch	E
Percina tanasi	Snail Darter	T
Phoxinus cumberlandensis	Blackside Dace	T
Plagopterus argentissimus	Woundfin	E, XN
Poeciliopsis occidentalis	Gila Topminnow (incl. Yaqui)	E
Prietella phreatophila	Mexican Blindcat (catfish)	E
Pristis clavata	Dwarf Sawfish	E
Pristis pectinata	Smalltooth Sawfish	E
Pristis pristis	Largetooth Sawfish	E
Pristis zijsron	Green Sawfish	E
Probarbus jullieni	Ikan Temoleh (minnow)	E
Ptychocheilus lucius	Colorado Pikeminnow (=squawfish)	E, XN
Rhinichthys osculus lethoporus	Independence Valley Speckled Dace	E
Rhinichthys osculus nevadensis	Ash Meadows Speckled Dace	E
Rhinichthys osculus oligoporus	Clover Valley Speckled Dace	E
Rhinichthys osculus ssp.	Foskett Speckled Dace	T
Rhinichthys osculus thermalis	Kendall Warm Springs Dace	E
Salmo platycephalus	Ala Balik (trout)	E
Salmo salar	Atlantic Salmon	E
Salvelinus confluentus	Bull Trout	XN, T
Scaphirhynchus albus	Pallid Sturgeon	E
Scaphirhynchus platorynchus	Shovelnose Sturgeon	S

Species	Common name	Status
Scaphirhynchus suttkusi	Alabama Sturgeon	E
Scleropages formosus	Asian Bonytongue	E
Sebastes paucispinis	Bocaccio	E
Sebastes pinniger	Canary Rockfish	T
Sebastes ruberrimus	Yelloweye Rockfish	T
Speoplatyrhinus poulsoni	Alabama Cavefish	E
Sphyrna lewini	Scalloped Hammerhead Shark	E, T
Spirinchus thaleichthys	Longfin Smelt (San Francisco Bay delta)	C
Tanakia tanago	Miyako Tango (=Toyko Bitterling)	E
Thaleichthys pacificus	Eulachon	T
Tiaroga cobitis	Loach Minnow	E
Xyrauchen texanus	Razorback Sucker	E

Appendix C: Methodology for Determining Replacement Costs of Fish

The replacement costs of fish presented in this publication are intended to update the costs last published in American Fisheries Society (AFS) Special Publication 30, printed in 2003. Those costs were based on a 2002 survey of state and private fish hatcheries. To provide comparability with the 2003 publication costs, a similar survey was developed with the aim of maximizing reliability while minimizing the respondent's burden of completing the survey. Data on the costs of producing fish on a per-inch basis, or, if not available, on a per-pound basis, were needed to comply with the AFS fish kill counting guidelines, as presented in this document and its predecessors (AFS 1982, 1992; Southwick and Loftus 2003). For private hatcheries, this meant collecting the prices charged per fish or per pound. Since government-operated hatcheries usually do not sell fish, the costs to raise fish in these hatcheries were collected by the survey.

A comprehensive list of hatcheries was developed to distribute the surveys. Included were private hatcheries identified in the 2003 survey, through trade publications, purchased business directories, and online searches. Contact information for federal and tribal hatcheries was provided by the U.S. Fish and Wildlife Service. State hatchery surveys were solicited through contacts with chief officer of the fisheries agency in each state.

Based on presurvey consultations with public hatchery managers, it was determined that hatcheries would be more receptive to surveys sent via e-mail. Prior to the survey, announcements were e-mailed to all possible state and federal hatchery managers and at regional and national conferences. The disposition of the surveys is as follows:

State hatcheries: Three rounds of surveys were e-mailed to 53 state fisheries agencies. Sixty-two surveys were returned from 34 states, including several states that completed separate surveys for each hatchery operation managed by the agency. Each nonresponding state was contacted regarding their survey. Most reported being unable to respond as the data needed were unavailable with a few unable to respond due to limited time or not having any hatcheries.

Federal hatcheries: Via three rounds of e-mailed surveys, 85 federal hatcheries were surveyed, with 22 completed surveys returned. Selected nonresponding federal hatcheries were contacted to inquire about their nonresponse, with all reporting either that not enough time to complete the survey or that the data were not available.

Private hatcheries: Via email, 165 private hatcheries identified in the 2003 study were contacted, including three reminders. Only four responded directly to the online survey. Separately, price lists for 238 private hatchery facilities were obtained from alternative sources, including phone calls and published lists. As with previous editions to this publication, pricing data are derived from public price lists provided by the facilities.

Tribal hatcheries: Fifteen tribal owned hatcheries were contacted via e-mail. Two hatcheries responded to the online survey.

Overall, 328 responses were received from fish production facilities, compared to 190 in the 2002 survey upon which the previous 2003 replacement costs were based.

Adjustments to Survey Data

When cost-per-inch data were not possible, cost-per-pound data are reported in the replacement costs tables. Few hatcheries regularly raise game species to trophy size, except for broodstock, nor do many raise species that are not recreationally important. Hatchery managers indicated that using the costs of raising and maintaining broodstock does not reflect the true costs one would incur when raising large fish for release. When possible, replacement costs for trophy-sized fish and for forage or bait fish are reported per pound.

The replacement cost tables present per-pound costs for larger fish. The per-pound costs were derived from the surveys' per-inch costs for the largest reported size of a species by using a formula that estimates the length-to-weight relationship of fish. Although exact length-to-weight ratios vary by species, season, age of fish, geographic location, and other factors, a general formula is useful in estimating either the fish length or weight. A general mathematical relationship between weight (W) and length (L) is (Carlander 1969, 1977)

$$W = cL^n$$

For most species, specific values for the constants c and n were obtained from sources assembled in Neumann et al. (2012), Murphy et al. (1991) or Schneider et al. (2000). Where a species-specific value was not available, values for a closely related species were substituted. In cases where the variability between price estimates at larger sizes was significantly greater than the variability at lower sizes, the price per pound was calculated using the lower (less variable) size and corresponding price as a basis.

The resulting per-pound costs for larger fish that are given in the replacement cost tables are conservative; that is, they do not adequately represent the costs of very large fish of any species. Since the values of c and n used in these calculations are averages derived from many studies, fishery managers should extrapolate the costs of large fish from local length–weight data, if these are available. It is important to note that per-pound estimates of larger fish may not accurately predict the per-pound costs for smaller fish since growth rates are typically slower for older fish. In other words, the per-pound estimates in this publication for larger fish may be higher than the per-pound estimate that one would expect for smaller fish, due to the cumulative cost of raising and maintaining larger fish over longer periods of time.

Size Increments

Hatcheries often did not report data for all sizes of a particular species. The variability of sizes reported by hatcheries would sometimes result in negative cost slopes. For example, the production cost average (calculated from the questionnaires) for a specific size increment would be lower than the cost for the next smaller size increment. Knowing it is not possible to raise an x-inch fish at a total cost lower than that for the size-class below it, it was necessary to replace the out-of-line production cost. In those cases, replacement cost observations were estimated by interpolating between two known price points that represented a generally upward sloping price curve.

Costs for Fish Not Reported by the Surveys

Species Represented

Hatchery responses were not received for all of the species reported herein. Efforts were made to identify a common trend in production costs for all species from to 2002 to 2016 but were largely unsuccessful. Cost increases (and decreases) varied by species and even per size ranges within a species. Therefore, for species not represented in the 2016 survey, the same methodology used by Southwick and Loftus (2003) to make such adjustments to costs were applied. Replacement costs for the species not reported in the 2016 hatchery survey were determined by applying the producer price index multiplier to the 2003 dollar costs. Where applicable, the per-pound costs for larger fish were calculated as described previously, with species-specific growth equations, where available, using these updated costs. The few generally marine-only species reported by hatcheries are presented in this publication if data were received for them. Other species being raised by hatcheries but not included here may be considered experimental and/or full-scale production cost data are unobtainable. Therefore, the costs of producing these species are not included. Striped Bass *Morone saxatilis*, Red Drum *Sciaenops ocellatus*, and Spotted Seatrout *Cynoscion nebulosus* costs are listed as those species frequent brackish waters where freshwater species are often present and may appear in fish kills in areas legally regarded as freshwater.

Public versus Private Hatcheries

Because private and public hatcheries calculate production costs differently, efforts were not made to equate the two. When fish are replaced after a fish kill, replacements can be obtained from either a public or private hatchery. Therefore, the averages of the two were used to determine replacement costs.

The private hatcheries' prices, or the cost of producing fish, generally include capital costs and presumably include a profit margin. Public hatchery cost data typically do not include capital costs, insurance, or taxes. Adjustments were only made to public hatchery surveys if capital costs were not included, as these are legitimate expenses incurred by agencies to produce and release fish, as explained previously.

Private and public hatcheries have differing costs because of differing missions. For example, public hatcheries respond to public needs regardless of a profit margin. Such efforts include preserving a particular strain of fish, reinforcing the populations of endangered or threatened species, tagging fish prior to release, and performing research on matters such as disease control. Private hatcheries are often free to choose whether to operate under these restraints and consequently tend to have a higher return for their fish production. These differences are pointed out to help explain the differences in cost, not to promote one over the other.

Appendix D: Legal Precedents

The concept of using fish replacement costs and economic values and/or procedures to calculate restitution for fish kills has been upheld in the following court cases. *This is not a complete list but constitutes examples of the application of American Fisheries Society (AFS) guidelines and/or values.*

Florida Department of Pollution Control v. International Paper Company
Florida Supreme Court, 1976, 329 So.2d 5

The state had developed tables of monetary values of fish. The court upheld that these tables were based on sufficient evidence and could be used to establish restitution for damage to fish.

City of Red Bank, Hamilton County, Tennessee v.
Tennessee Water Quality Control Board
Chancery Court of Davidson County, Tennessee, July 22, 1976

The commissioner of the Tennessee Department of Public Health assessed the city of Red Bank $3,304.79 for fish killed.

The city of Red Bank appealed, charging, among other things, that the method of determining the number of dead fish and their cost was invalid. The court concluded that the methods of counting and evaluating the fish killed had a rational basis and should not be overturned. The court also stated that the conclusions of the board were based on substantial evidence and should be sustained.

Virginia ex. Rel. Water Control Board v. Weaver Mirror Company
Circuit Court, Franklin County, Virginia, June 2, 1977

The circuit court returned a verdict awarding the state $10,043 for the monetary costs of fish killed.

Defendants challenged the monetary costs and the fish counting procedure on hearsay and validity grounds. The judge overruled the objections and held both the assessed monetary costs and counting guidelines to be valid.

A. F. Chastain v. Tennessee Water Quality Control Board
Supreme Court of Tennessee, 1977, 555 S.W.2d 113

The Tennessee Supreme Court affirmed decree of the Chancery Court, Davidson County, Tennessee, which affirmed an award of $2,676.13 assessed by the commissioner of the Tennessee Department of Public Health for a pollution-caused fish kill.

City of Murray v. Kentucky Department of Fish and Wildlife
Commonwealth of Kentucky Court of Appeals, Calloway County, May 11, 1978

The court of appeals upheld a lower court decision that accepted the use of the fish kill counting guidelines (AFS 1982).

State of Oklahoma v. Kerr-McGee Corp.
Supreme Court of Oklahoma, 1980, 619 P.2d 858

The court upheld that the costs of restocking the site of a fish kill was a valid method of assessing damages. Expert testimony as to the cost of restocking was supported by sufficient competent evidence that the amount fined was the appropriate restitution cost.

Constantine v. State of Alaska
Alaska Appellate Court, 1987, 739 P.2d 188

The court upheld a state statute that a fisheries violator could be assessed a fine per violation plus a fine "not less than the gross value to the fishermen of fish found on the vessel or at the fishing site at the time of the violation."

Wilson v. State of Alaska
Alaska Appellate Court, 1988, 756 P.2d 307

The court ruled that the "minimum fine for fishing violation, where fish are found, is the value of the fish, but where value of the fish exceeds $10,000, the minimum penalty and the maximum penalty for the first offenders is the value of the fish.

Indiana Department of Natural Resources v. Clay Furness et al.
Indiana Appellate Court, 1989, 546 N.E.2d 322

The court ruled in favor of the Department of Natural Resources (DNR), who sought restitution for salmon and trout caught incidentally in the Lake Michigan perch fishery. The salmon and trout were originally stocked by the DNR when the fish were 3.5 in long. When the fish were netted, they had grown to lengths of 9–13 in. A lower court first ruled that restitution for incidentally caught salmon and trout should be based on the original stocking costs. However, the state supreme court overruled, saying replacement costs should be based on the size of the fish killed rather than on the size of the fish when first stocked. Restitution was assigned according to costs in AFS Special Publication 13 (AFS 1982).

Indiana Department of Natural Resources v. Jason Sweet, Neil Sweet, and Donald Sweet
State of Indiana, Circuit Court of Wabash County, 2012, Case No. 85C01-1207-PL-650

This court case in Indiana upheld AFS values when designating the monetary value of the fish kill, but the Indiana statute 14-22-10-6 did not authorize the Indiana Department of Natural Resourcses to collect investigative costs. The statute has since been rewritten to include investigative costs.

Monroe Branstad v. State of Iowa Natural Resource Commission, and the Iowa Department of Natural Resources

Supreme Court of Iowa, 2015, Case No. 14-0205

In this series of court cases (the final case listed above), the Iowa DNR sought restitution for a large fish kill resulting from sweet corn silage runoff into a nearby river amounting to $61,741 in damages. Branstad appealed, and after judicial review, the district court found that the Iowa DNR did not follow AFS guidelines correctly and consistently when calculating an assessment of the amount fish killed. Therefore, the restitution was calculated from only the dead fish that were actually counted, and the restitution was lowered to $5,298.

Edward Mosely v. State of Missouri Clean Water Commission

Circuit Court of Moniteau County, Missouri, March 6, 1990

The Missouri Clean Water Commission assessed Mr. Mosely $5,631.25 for fish killed plus civil fines and investigation costs.

State of Missouri and Missouri Department of Natural Resources v. Tyson Foods

Circuit Court of Barry County, Missouri, 2015, Case No. 14BR-CC00071

The court upheld AFS guidelines when it ruled for Tyson Farms to pay $162,898.78 in damages to aquatic life and its habitat due to polluting a creek in Missouri. Tyson Farms was also compelled to pay investigation costs of the fish kill to Missouri's DNR and its Department of Conservation totaling $11,075.17 and $36,026.05, respectively. All restitution was based on AFS guidelines.

State of Missouri and Missouri Department of Natural Resources v. Ozark Mountain Technologies

Circuit Court of Crawford County, Missouri, 2015, Case No. 15CF-CC00018

The circuit court upheld AFS guidelines when assigning the monetary values of the fish kill and investigative costs to Ozark Mountain Technologies.

State of Missouri and Missouri Department of Natural Resources v. Bernard Bargfrede

Circuit Court of Saline County, Missouri, 2015, Case No. 13SA-CV00851

The court ruled that Bernard Bargfrede would pay the DNR's investigation costs of the fish kill according to AFS guidelines, as well as a civil penalty of $1,500.

State of Kansas

The following is a list of cases where the state successfully prosecuted instances of illegal commercialization of fish per Kansas statute 32-1005 (8):

County	Case name	Case number	Year
Sedgwick and Saline	State of Kansas v. Thanh Thuy Oriental Food Store	#92-213	1994
Sedgwick	State of Kansas v. Kim Nhung Market	#92-214	1994
Saline	State of Kansas v. Martin Minh	#93-219	1994
Johnson	State of Kansas v. Makong Market	#93-231	1993
Johnson	State of Kansas v. Asian Food Market	#93-232	1993
Geary	State of Kansas v. Seoul Oriental Food Market	#94-236	1994
Geary	State of Kansas v. First Oriental Food Gift Market	#94-237	1994
Clay	State of Kansas v. Lyman Crawford	#94-239	1995
Clay	State of Kansas v. Matthew McGinty	#94-239	1995
Shawnee	State of Kansas v. Fidel Adel-Alim	#96-264	1998
Ellsworth Marion	State of Kansas v. Troy Russell	#99-309	1999
McPherson	State of Kansas v. J. D. Starnes	#99-310	1999
Shawnee	State of Kansas v. Renne Shumway	#99-325	2001

Appendix E: Federally Listed Endangered and Threatened Freshwater Mollusks[1]

Summarized below are mussel taxa that are included, or are being reviewed for inclusion, in the list of endangered and threatened wildlife under the U.S. Endangered Species Act of 1973. This summary was compiled from the U.S. Fish and Wildlife Service's (USFWS) Endangered Species database accessed on July 31, 2017 (www.fws.gov/endangered). Readers should check this source for any changes or updates to the list. State and provincial governments also have statutes protecting rare, threatened, or endangers species and should also be consulted for a list of threatened and endangered species and associated regulations.

The scientific names and common names are as listed by the U.S. Fish and Wildlife Service. All species are protected wherever they occur unless noted in the table footnotes for "Federal listing status." Listing status codes used are

E = Endangered
T = Threatened
C = Candidate for listing
P = Proposed for listing
XN = Experimental population, nonessential

Scientific name	Common name	Federal listing status
Alasmidonta atropurpurea	Cumberland elktoe	E
Alasmidonta heterodon	Dwarf wedgemussel	E
Alasmidonta raveneliana	Appalachian elktoe	E
Amblema neislerii	Fat threeridge (mussel)	E
Arkansia wheeleri	Ouachita rock pocketbook	E
Cumberlandia monodonta	Spectaclecase (mussel)	E
Cyprogenia stegaria	Fanshell	E, XN[a]
Cyrtonaias tampicoensis tecomatensis	Tampico pearlymussel	E
Dromus dromas	Dromedary pearlymussel	E[b], XN[a, c]
Elliptio chipolaensis	Chipola slabshell	T
Elliptio lanceolata	Yellow lance	P
Elliptio spinosa	Altamaha spinymussel	E
Elliptio steinstansana	Tar River spinymussel	E
Elliptoideus sloatianus	Purple bankclimber (mussel)	T

[1] This list was updated at the time of publication. Please visit the USFWS Web site (www.fws.gov/endangered/species/us-species.html) for an updated list of federally threatened and endangered species and contact state fish and wildlife agencies for a list of threatened and endangered species and associated regulations.

Scientific name	Common name	Federal listing status
Epioblasma brevidens	Cumberlandian combshell	E[b], XN[a, c]
Epioblasma capsaeformis	Oyster mussel	E[b], XN[a, c]
Epioblasma florentina curtisii	Curtis pearlymussel	E
Epioblasma florentina florentina	Yellow blossom (pearlymussel)	E[b], XN[c]
Epioblasma florentina walkeri (=*E. walkeri*)	Tan riffleshell	E
Epioblasma metastriata	Upland combshell	E
Epioblasma obliquata obliquata	Purple cat's paw (=purple cat's paw pearlymussel)	E[b], XN[c]
Epioblasma obliquata perobliqua	White catspaw (pearlymussel)	E
Epioblasma othcaloogensis	Southern acornshell	E
Epioblasma penita	Southern combshell	E
Epioblasma torulosa gubernaculum	Green blossom (pearlymussel)	E
Epioblasma torulosa rangiana	Northern riffleshell	E
Epioblasma torulosa torulosa	Tubercled blossom (pearlymussel)	E[b], XN[c]
Epioblasma triquetra	Snuffbox mussel	E
Epioblasma turgidula	Turgid blossom (pearlymussel)	E[b], XN[c]
Fusconaia burkei	Tapered pigtoe	T
Fusconaia cor	Shiny pigtoe	E[b], XN[a, c]
Fusconaia cuneolus	Finerayed pigtoe	E[b], XN[a, c]
Fusconaia escambia	Narrow pigtoe	T
Fusconaia rotulata	Round Ebonyshell	E
Hamiota australis	Southern sandshell	T
Hemistena lata	Cracking pearlymussel	E[b], XN[a, c]
Lampsilis abrupta	Pink mucket (pearlymussel)	E
Lampsilis altilis	Finelined pocketbook	T
Lampsilis bracteata	Texas fatmucket	C
Lampsilis higginsii	Higgins eye (pearlymussel)	E
Lampsilis perovalis	Orangenacre mucket	T
Lampsilis powellii	Arkansas fatmucket	T
Lampsilis rafinesqueana	Neosho Mucket	E
Lampsilis streckeri	Speckled pocketbook	E
Lampsilis subangulata	Shinyrayed pocketbook	E
Lampsilis virescens	Alabama lampmussel	E[b], XN[c]
Lasmigona decorata	Carolina heelsplitter	E
Lemiox rimosus	Birdwing pearlymussel	E[b], XN[a, c]
Leptodea leptodon	Scaleshell mussel	E
Margaritifera hembeli	Louisiana pearlshell	T
Margaritifera marrianae	Alabama pearlshell	E
Medionidus acutissimus	Alabama moccasinshell	T

Scientific name	Common name	Federal listing status
Medionidus parvulus	Coosa moccasinshell	E
Medionidus penicillatus	Gulf moccasinshell	E
Medionidus simpsonianus	Ochlockonee moccasinshell	E
Medionidus walkeri	Suwannee moccasinshell	T
Megalonaias nicklineana	Nicklin's pearlymussel	E
Obovaria retusa	Ring pink (mussel)	E, XN[a]
Pegias fabula	Littlewing pearlymussel	E
Plethobasus cicatricosus	White wartyback (pearlymussel)	E, XN[a]
Plethobasus cooperianus	Orangefoot pimpleback (pearlymussel)	E, XN[a]
Plethobasus cyphyus	Sheepnose mussel	E
Pleurobema clava	Clubshell	E[b], XN[c]
Pleurobema collina	James spinymussel	E
Pleurobema curtum	Black clubshell	E
Pleurobema decisum	Southern clubshell	E
Pleurobema furvum	Dark pigtoe	E
Pleurobema georgianum	Southern pigtoe	E
Pleurobema gibberum	Cumberland pigtoe	E
Pleurobema hanleyianum	Georgia pigtoe	E
Pleurobema marshalli	Flat pigtoe	E
Pleurobema perovatum	Ovate clubshell	E
Pleurobema plenum	Rough pigtoe	E, XN[a]
Pleurobema pyriforme	Oval pigtoe	E
Pleurobema strodeanum	Fuzzy pigtoe	T
Pleurobema taitianum	Heavy pigtoe	E
Pleuronaia dolabelloides	Slabside pearlymussel	E
Popenaias popeii	Texas hornshell	PE
Potamilus capax	Fat pocketbook	E
Potamilus inflatus	Alabama (=inflated) heelsplitter	T
Ptychobranchus greenii	Triangular kidneyshell	E
Ptychobranchus jonesi	Southern kidneyshell	E
Ptychobranchus subtentum	Fluted kidneyshell	E
Quadrula cylindrica cylindrica	Rabbitsfoot	T
Quadrula cylindrica strigillata	Rough rabbitsfoot	E
Quadrula fragosa	Winged mapleleaf	E[b], XN
Quadrula houstonensis	Smooth pimpleback	C
Quadrula intermedia	Cumberland monkeyface (pearlymussel)	E[b], XN[a, c]
Quadrula petrina	Texas pimpleback	C

Scientific name	Common name	Federal listing status
Quadrula sparsa	Appalachian monkeyface (pearlymussel)	E, XN[a]
Quadrula stapes	Stirrupshell	E
Toxolasma cylindrellus	Pale lilliput (pearlymussel)	E
Truncilla macrodon	Texas Fawnsfoot	C
Villosa choctawensis	Choctaw bean	E
Villosa fabalis	Rayed Bean	E
Villosa perpurpurea	Purple bean	E
Villosa trabalis	Cumberland bean (pearlymussel)	E[b], XN[a, c]

[a] Portions of the French Broad and Holston Rivers, TN.

[b] Wherever found (except Experimental Populations).

[c] Portions of Tennessee River, AL.

Appendix F: Hatchery Replacement Costs of Freshwater Mussels at Taggable Size

Genera of freshwater mussels are based on Williams et al. (2017). Federal threatened and endangered designations are updated through July 2017. *These guidelines should not be used to assign replacement cost for threatened and endangered species.* For all mussel kills involving federal threatened and endangered species, regional U.S. Fish and Wildlife offices should be notified as federal regulations take precedence in such situations.

Cost categories for production of mussels are based on grouping of genera with similar requirements and difficulty of cultivation. The costs reflect production of mussels of taggable size, which can vary by genus and species.[1] Costs within a grouping of genera are based on a weighted average cost of producing individual species within that group of genera. The weighting applied to the individual observations were developed by malacologists with knowledge of mussel cultivation practices to account for varying volumes and maturity of production methods across the reporting facilities.

The costs below are not directly comparable to those published in the 2003 edition due to key conceptual and methodological differences. The 2003 publication presented replacement costs of juvenile mussels at a standard age of 2 months. At that age, the overall survival rate to 5 years was estimated at 9.5%. Since most mussels are not able to be tagged at that age due to size, the present study collected information on the cost to raise mussels to a taggable size. The age of mussels of taggable size ranges from 6 months to 5 years, with an average of 2 years, resulting in higher culture costs per mussel. Conversely, the survival rate of mussels at this size is higher requiring fewer replacement mussels to achieve a desired population after 5 years.

A genus designated with an asterisk indicates that all species are fully federally listed. The two genera designated as "NA" are unknown and unlike other mussel genera in North America. If animals have to be replaced, significant additional research will be required and investigators should defer to malacologists knowledgeable in the life history of that species to identify a comparable genus in terms of cultivation requirements and difficulty.

[1] AFS Special Publication 30 (2003) presented costs of juvenile mussels at 2 months of age, which is different from taggable size for most species. Costs in this publication are not comparable to the earlier cost estimates.

Genus	Cost per mussel	Genus	Cost per mussel
Actinonaias	$25.97	*Medionidus*	$28.00
Alasmidonta	$92.16	*Megalonaias*	$129.30
Amblema	$101.37	*Obovaria*	$32.29
Anodonta	$34.01	*Obliquaria*	$47.50
Anodontoides	$34.01	*Parvaspina*	$59.64
Arcidens	$70.00	*Pegias*	*
Cyclonaias	$129.30	*Plectomerus*	$47.50
Cyprogenia	*	*Plethobasus*	*
Cyrtonaias	*	*Pleurobema*	$60.99
Disconaias	*	*Pleuronaia*	$60.99
Dromus	*	*Popenais*	*
Ellipsaria	$77.72	*Potamilus*	$47.50
Elliptio	$50.90	*Ptychobranchus*	$30.00
Elliptoideus	*	*Pyganodon*	$34.01
Epioblasma	*	*Quadrula*	$129.30
Eurynia	$50.90	*Reginaia*	$60.99
Fusconaia	$60.99	*Simpsonaias*	$212.48
Hamiota	$27.20	*Sinanodonta*	NA
Hemistena	*	*Strophitus*	$25.00
Glebula	$27.20	*Theliderma*	$129.30
Gonidea	NA	*Toxolasma*	$50.00
Lampsilis	$27.20	*Tritigonia*	$129.30
Lasmigona	$62.72	*Truncilla*	$82.30
Lemiox	*	*Uniomeris*	$129.30
Leptodea	$22.93	*Utterbackia*	$34.01
Ligumia	$30.68	*Utterbackiana*	$34.01
Margaritifera	$70.00	*Venustaconcha*	$67.71
		Villosa	$67.71

Appendix G: Random Numbers[1]

48461	14952	72619	73689	52059	37086	60050	86192	67049	64739
76534	38149	49692	31366	52093	15422	20498	33901	10319	43397
70437	25861	38504	14752	23757	59660	67844	78815	23758	86814
59584	03370	42806	11393	71722	93804	09095	07856	55589	46020
04285	58554	16085	51555	27501	73883	33427	33343	45507	50063
77340	10412	69189	85171	29082	44785	83638	02583	96483	76553
59183	62687	91778	80354	23512	97219	65921	02035	59847	91403
91800	04281	39979	03927	82564	28777	59049	97532	54540	79472
12066	24817	81099	48940	69554	55925	48379	12866	51232	21580
69907	91751	53512	23748	65906	91385	84983	27915	48491	91068
80467	04873	54053	25955	48518	13815	37707	68687	15570	08890
78057	67835	28302	45048	56761	97725	58438	91528	24645	18544
05648	39387	78191	88415	60269	94880	58812	42931	71898	61534
22304	39246	01350	99451	61862	78688	30339	60222	74052	25740
61346	50269	67005	40442	33100	16742	61640	21046	31909	72641
66793	37696	27965	30459	91011	51426	31006	77468	61029	57108
86411	48809	36698	42453	83061	43769	39948	87031	30767	13953
62098	12825	81744	28882	27369	88183	65846	92545	09065	22655
68755	06261	54265	16203	23340	84750	16317	88686	86842	00879
52679	19595	13687	74872	89181	01939	18447	10787	76246	80072
84096	87152	20719	25215	04349	54434	72344	93008	83282	31670
63964	55937	21417	49944	38356	98404	14850	17994	17161	98981
31191	75131	72386	11689	95727	05414	88727	45583	22568	77700
30545	68523	29850	67833	05622	89975	79042	27142	99257	32349
52573	91001	52315	26430	54175	30122	31796	98842	37600	26025
16586	81842	01076	99414	31574	94719	34656	80018	86988	79234
81841	88481	61191	25013	30272	23388	22463	65774	10029	58376
43563	66829	72838	08074	57080	15446	11034	98143	74989	26885
19945	84193	57581	77252	85604	45412	43556	27518	90572	00563
79374	23796	16919	99691	80276	32818	62953	78831	54395	30705
48503	26615	43980	09810	38289	66679	73799	48418	12647	40044
32049	65541	37937	41105	70106	89706	40829	40789	59547	00783
18547	71562	95493	34112	76895	46766	96395	31718	48302	45893
03180	96742	61486	43305	34183	99605	67803	13491	09243	29557
94822	24738	67749	83748	59799	25210	31093	62925	72061	69991

[1] From Rohlf and Sokal (1969).